Abusing Science

Abusing Science

The Case against Creationism

Philip Kitcher

The MIT Press
Cambridge, Massachusetts
London, England

Printed and bound in the United States of America.

Library of Congress Cataloging in Publication Data

Kitcher, Philip, 1947–
  Abusing science.

  Bibliography: p.
  Includes index.
  1. Evolution. 2. Evolution and religion. 3. Creation. I. Title.
QH371.K57   1982   575.01   82–9912
                              AACR2
ISBN 978-0-262-11085-3 (hc.: alk. paper) 978-0-262-61037-7 (pb.: alk. paper)

20 19 18 17 16

For Penny Conti
and dedicated teachers everywhere

Contents

Preface

Many people have helped in writing this book. Without their advice, encouragement, criticisms, and suggestions it would have been very different—or it would not have existed at all. I want to begin with some words of thanks for the generous contributions of friends and colleagues.

My chief debt is to Patricia Kitcher, with whose help the book was written and without whose help it could not have been written. She gave up six weeks from her own research to work over and to rewrite drafts. She reorganized some sections, rewrote others, and constantly supplied me with advice and criticisms. She is coauthor of the last chapter; her contributions to it were even more extensive than those she made to other parts of the book.

At an early stage of my research I received helpful suggestions from a number of people. Stephen Brush, David Raup, Robert Richards, David Schramm, and especially William Wimsatt all offered valuable advice. More recently, I have benefited from the constructive criticisms of Steven Cahn, Peter Demetz, Rebecca German, Norman Gilinsky, Hilary Kornblith, William Mann, Peter Williamson, and Kurt Wise.

There are six people who deserve special thanks for their careful and detailed responses to the penultimate draft: John Beatty, Robert Dorit, Clark Glymour, Stephen Jay Gould, Richard Lewontin, and Gregory Mayer. I appreciate greatly the time that these people spent in helping me, and I hope they will think that it was worthwhile.

I would never have had the opportunity to take advantage of so much good advice had it not been for the expertise of Mrs. Leslie Weiger. She typed the penultimate draft in time for me to circulate it and, undaunted, went on to type the final version. Her speed, efficiency, kindness, and enthusiasm are much appreciated.

Finally, my thanks to the Museum of Comparative Zoology at Harvard University for its warm hospitality during the time that this book was written.

February 1982
Belmont, Massachusetts

The Creationist Crusade

In 1860, at the meetings of the British Association in Oxford, Thomas Henry Huxley, "Darwin's bulldog," vanquished Bishop Wilberforce in a famous debate. Charles Darwin had published *The Origin of Species* the year before. His book contained very little about human evolution; Darwin had stepped cautiously around the issue. Still, Wilberforce saw the implications of Darwin's views. A skilled debater, whose slippery performances had earned him the nickname of "Soapy Sam," he thought he saw a way to achieve rhetorical effect. In an unwise moment, toward the end of his address, he turned to Huxley, who was sitting beside him on the platform. With the air of a man about to deliver the fatal blow, he asked whether Huxley's descent from the apes came through his grandfather or his grandmother. Popular history reports that Huxley muttered under his breath, "The Lord has delivered him into my hands." He went on to deliver a scathing response, openly admitting that he would prefer an ape for a grandparent to a man, blessed with intellect and education, who used rhetorical tricks to confuse an important scientific issue.

Over 120 years later, the conclusions and debating methods of Soapy Sam are alive and well and playing in Peoria. In recent years, a political alliance has been forged between the self-appointed champions of virtue and religion—the Moral Majority—and a group of believers in the literal truth of the Bible. These extreme fundamentalists, who call themselves Scientific Creationists, have founded the Institute for Creation Research. Many of them had obtained doctorates in various scientific fields, but their energies are now channelled into promoting the Creationist cause. They have written and distributed a large mass of literature, arguing that evolutionary theory suffers from crucial deficiencies and that Creation Science, a doctrine compatible with the literal truth of Genesis, is far better supported by the evidence. The titles of their books convey the message: *The Natural*

Sciences Know Nothing of Evolution; Creation: A Scientist's Choice; That You Might Believe, The Bible Has the Answer; Scientific Creationism; The Troubled Waters of Evolution; Evolution? The Fossils Say No!; Evolution: Its Collapse in View?; The Great Brain Robbery; and many more in the same vein.

The goals of the alliance are formidable. The Moral Majority and the Institute for Creation Research would like to "reform" the teaching of high school science. Creationists have already achieved some dramatic successes. In seventeen states legislators have introduced bills that would require the teaching of Creation "science." Two states, Arkansas and Louisiana, have passed "balanced-treatment" laws. These legislative victories are among the most prominent achievements of the Moral Majority. They encourage the faithful to believe that the cause is being advanced.

The alliance between the Moral Majority and the "scientific" Creationists serves both allies well. Jerry Falwell's Old Time Gospel Hour offers a forum for broadcasting Creationist ideas. On the other hand, Falwell needs concrete issues around which to build his movement. Those who are drawn in are exhorted to carry the cause into their local schools. Tactical advice is readily available. Readers of the Creationist newsletter are encouraged to make lists of teachers who promote evolutionary theory or argue against Creationism. And, for a small fee, the Institute for Creation Research will send out a pamphlet, written by its director, Henry Morris. In "Introducing Creationism in the Public Schools," Morris tells parents and students how they can help. He suggests that concerned parents might "establish a formal community organization, with some appropriate name (Citizens for Scientific Creationism, Parents Concerned for Educational Integrity, Civil Rights for Creationists, Committee for the Improvement of Education, etc.)"; or they could poll the views of their community; students who believe in Creationism are encouraged to raise questions in class — but they are urged to be respectful and courteous, "winsome and tactful, kind and patient"; Morris also adds that "cleanliness and neatness don't hurt either" (Morris 1975, 7–8).* So the foot soldiers of the cause, well-heeled and well-scrubbed, are set in motion.

It is easy to overlook these smaller skirmishes. For most people, the image of the battle between Creationism and evolution is that

*Works are cited in text in parentheses by author, date of publication, and page(s). For fuller bibliographical information (title of work, publisher, place of publication), please consult the list of references. The great advantage of this system of citation is that it dramatically reduces the number of notes. I hope that people who are unfamiliar with it will quickly feel at home—and will enjoy not having to keep one finger marking the place where notes are found.

presented in a deservedly famous movie, *Inherit the Wind*. The image proves comforting. All will be well, for a latter-day Clarence Darrow—or, better, a latter-day Spencer Tracy, playing Clarence Darrow—will come forth to achieve a moral victory for science. But the popular image of the Scopes trial as a defeat for fundamentalism is inaccurate. Though Spencer Tracy won the hearts of moviegoers, Darrow failed to win the day for evolution. Scopes was convicted. (Rightly so. He had broken a Tennessee law forbidding the teaching of evolution.) His conviction was overturned on a technicality. As a result, the case did not go forward to appeal, and it was therefore impossible to challenge the constitutionality of the law. (Antievolution laws survived in Arkansas, Mississippi, and Tennessee until the 1960s.) For all Darrow's clever questions, the champions of science did not gain their point.

Indeed, the legacy of the Scopes trial reveals the extent of the defeat. Biology textbooks were changed to soft-pedal the teaching of evolution. For about thirty years, until the launching of Sputnik led many people to think that science education should be taken seriously, publishers and educators were persuaded to follow the educational policy that some fundamentalists had tried to require by law. The moral of history is echoed in recent events. On the day that Judge William Overton, in a lucid and informed ruling, struck down Arkansas law 590 (the Arkansas balanced-treatment law), the Mississippi senate voted, 48–4, to introduce its own balanced-treatment law.

Even if Creationists continue to lose in the courts, they may still succeed in wreaking havoc upon science education (and, ultimately, upon American science). By lobbying local school administrators, the Creationist minions can affect the books that are chosen and the curriculum that is designed. Because textbooks are published to make a profit, the special-interest pressure will change the character of the books that are produced. While Creationist laws fail, the cause may triumph, as science education relapses into its post-Scopes, pre-Sputnik condition.

When the Arkansas Creationist law was challenged, a team of scientific, philosophical, and theological luminaries assembled in Little Rock. Judge Overton drew on expert testimony. Teachers, administrators, and local school-board members are likely to be less lucky. The advocates of Creationism may be winsome, tactful, and sweetly reasonable, but they are bent on having their way and they are equipped with the ideas and arguments of the Institute for Creation Research. Ready with the rhetorical tricks of Creationist literature, a carefully primed student or a concerned parent can easily embarrass a teacher or a PTA member. Not every school district has its Huxley, prepared to respond to the clever questions of the local Wilberforce.

Although it attempts to do more, this book is intended to be a manual for intellectual self-defense, something that can be consulted when the smiling advocates of Creationism launch their attack. I hope that it will help anyone who wants to arrive at an informed opinion on the issue. It is written for concerned citizens, whether their background in science is minimal or they are professional scientists. I have tried to explain what the Creationists say and why they are wrong—in such a way that anyone who is interested may find out.

The chapters that follow attempt to disentangle and address the major arguments that "scientific" Creationists have given. Creationist strategy is often to run entirely different issues together, to concoct a muddy paste out of distinct allegations about the evils of evolution and the glories of Creationism. Falwell's printed letter, soliciting (tax-deductible) contributions from all those who request information about the Moral Majority's position on the issue, is a masterpiece of the art. In one and a half pages, Falwell charges that belief in evolution requires more faith than belief in special creation, that "academic freedom" requires teaching both sides of the story, that the teaching of evolution fosters the attitude that we may as well "eat, drink, and be merry." The questions Falwell raises need to be treated separately: What are the respective scientific credentials of evolutionary theory and Creation "science"? What does a genuine intellectual tolerance require of us? What consequences does evolutionary theory have for our moral views and our moral practices? The book that follows is a chase. The Creationist is allowed to choose one battleground after another. Given each choice of battleground, I insist that the battle be fought on that ground. In every case, "scientific" Creationism is defeated. When all the distortions have been removed, all the attempts to flaunt credentials examined, all the misleading quotations returned to their contexts, all the fallacies laid bare, we shall see Creation "science" for what it is—an abuse of science.

One important theme that I shall emphasize is that, although the Creationist campaign is advertised as an assault on evolutionary theory, it really constitutes an attack on the whole of science. Evolutionary biology is intertwined with other sciences, ranging from nuclear physics and astronomy to molecular biology and geology. If evolutionary biology is to be dismissed, then the fundamental principles of other sciences will have to be excised. All other major fields of science will have to be trimmed—or, more exactly, mutilated—to fit the Creationists' bill. Moreover, in attacking the methods of evolutionary biology, Creationists are actually criticizing methods that are used throughout science. As I shall argue extensively, there is no basis for separating the procedures and practices of evolutionary biology from those that are fundamental to all sciences. If we let the Creationists have their way, we may as well go whole hog. Let us reintroduce the flat-earth theory, the chemistry of the four elements, and mediaeval astrology. For these outworn doctrines have just as much claim to rival current scientific views as Creationism does to challenge evolutionary biology.

My treatment of the methods of science indicates one way in which the book aspires to do more than help the beleaguered victims of Creationist assaults. By training, I am a philosopher of science, a person whose business consists in trying to understand what science is and how it works. Ironically, philosophers of science owe the Creationists a debt. For the "scientific" Creationists have constructed a glorious fake, which we can use to illustrate the differences between science and pseudoscience. By examining their scientific pretensions, I have tried to convey a sense of the nature and methods of science. I hope that this book will correct some popular misapprehensions about science, and that it will offer a picture that working scientists will find both congenial and useful.

A brief outline may help readers to find quickly the discussions that are most relevant to their interests. I begin with a brief introduction to evolutionary theory. As anyone who has dipped into the Creationist literature will have realized, Creationists have some acquaintance with scientific terminology; they often use technical language to present their arguments. Successful self-defense requires a modest literacy in the language of evolutionary theory. My first chapter attempts to convey just that. People who are familiar with contemporary evolutionary theory will want to skip this chapter and plunge in to the controversy. Others may also want to begin with chapter 2 and turn back to chapter 1 only when they encounter unfamiliar ideas.

Chapters 2–4 constitute a defense of evolutionary theory against Creationist objections. I start with the most global criticisms. Creationists delight in laying down criteria for science and then arguing that evolutionary theory does not meet their criteria. After showing that their criteria are based on a misunderstanding of the scientific enterprise, I point out that evolutionary theory not only passes the real tests for successful science but that it does so with flying colors. Chapter 3 moves on to consider more specific Creationist complaints about the methods of evolutionary biology. These turn out to be based on misunderstanding of the methods of science or of evolutionary theory—

or of both. In chapter 4 I take up the "scientific facts" that are supposed to refute evolution. All of the Creationists' favorite examples find their place: from the most global to the most local; from the appeal to thermodynamics, through the nature of the fossil record, to very particular biological "findings."

The fifth chapter offers a critical evaluation of Creation "science." After unearthing what there is of a positive doctrine in the Creationists' welter of words, I consider whether the doctrine can be used to solve any scientific problems. In this context, I discuss the significance of recent debates within evolutionary theory and the Creationist proposals for revising the geological time scale. Chapter 6 turns to educational questions, asking in particular what genuine intellectual tolerance requires of us. That chapter concludes with a brief look at some of the tactics that Creationists use in exploiting tolerance. The seventh (and last) chapter examines the charge that evolutionary theory is intolerable because it is inimical to religion and morality. Of course, this is what the fuss is all about. However, it will turn out that the theory of evolution does not predict (or encourage) the extinction of morality. Nor is the theory opposed to religious belief in general, but only to the views of a particular sect—Creation "scientists."

This book is not, therefore, an attempt to debunk religion. Nor does it criticize those who accept a literal reading of the Genesis account of Creation simply as an article of religious faith. Although I am critical of Creationism, my business is strictly with a political movement, the movement launched by the "scientific" Creationists and their friends in the Moral Majority. The scientific evidence tells decisively against the literal truth of Genesis. That fact does not mean that religion is refuted. Nor should it perturb anyone who believes, in the tradition of Tertullian and Kierkegaard, that faith can and should transcend any scientific findings. I quarrel only with those who pretend that there is scientific evidence to favor the Genesis story understood literally, who masquerade religious doctrines as scientific explanations, and who try to persuade their fellow citizens to make religious teaching a part of education in science.

1

Evolution for Everyone

The origin

"With such moderate abilities as I possess, it is truly surprising that thus I should have influenced to a considerable extent the beliefs of scientific men on some important points." So ends Darwin's *Autobiography* (Darwin 1876/Barlow 1969). Darwin was far too modest. The more that historians have studied his remarkable career, the more apparent it has become that Darwin was a man of exceptional talents, that he was blessed with a rare combination of deep insight, broad vision, and a careful and patient eye for details. What is truly surprising is that even educated and intelligent people today know only a caricature of Darwin's ideas. The central doctrines of evolutionary theory are still swathed in misconceptions. As H. J. Muller exclaimed on the centennial of *The Origin of Species*, "One hundred years without Darwin is enough!"

My overview of evolutionary theory will begin with Darwin and with his important contributions. The main thesis of evolution is that species are not fixed and immutable. One kind of organism can have descendants that belong to a different kind. From one original species, a number of different kinds may be generated. Evolutionary biologists believe that the birds are all descendants of a particular kind of reptile and that both cats and dogs have come from a common mammalian stock.

Darwin was not the first evolutionist. At the beginning of the nineteenth century, the French biologist Jean Baptiste de Lamarck proposed that species may take on new forms in response to their needs. Lamarck's ideas are more subtle and less definite than they are often portrayed as being. However, a hoary example may serve to distinguish his views from Darwin's. Lamarck would explain the giraffe's long neck as follows. Primitive, short-necked giraffes would have been unable to browse on the leaves of tall trees. Driven by its need for food, each individual primitive giraffe stretched its neck. The giraffes of the next generation benefited from this communal stretching. They inherited the characteristics acquired by their industrious parents. In their turn, they too reached upward. The result was a sequence of giraffes with ever longer necks, a sequence that culminates in the modern form.

This style of evolutionary explanation is not Darwin's. According to Darwin, the principal mechanism for evolution is *natural selection*. Like the plant or animal breeder, nature selects. Pigeon fanciers (whose doings Darwin studied carefully) choose to breed only those birds with the features that particularly interest them. Analogously, nature "chooses" for survival and reproduction those organisms whose characteristics have best equipped them to compete in a struggle for limited resources. In almost any natural population of organisms, more offspring will be produced than are able to survive. The offspring will vary—in particular, they will vary with respect to characteristics that affect their abilities to survive and reproduce. Some organisms will survive longer and reproduce more frequently. If the advantageous characteristics are inheritable, then they will be transferred to descendants. As a result, they will become more prevalent in later generations. Over a large number of generations the common features of the population may be radically changed.

The idea of evolution by natural selection will be clearer if we contrast Lamarck's giraffe with Darwin's. The Darwinian explanation of the evolution of the giraffe would begin with some initial group of short-necked giraffes (the *ancestral population*). Though all the giraffes in this group had short necks, some happened to have longer necks than others. These more fortunate beasts were able to browse on foliage that their fellows could not reach. With greater opportunities to feed well, they were better able to survive and multiply. So, in the next generation, the frequency of giraffes with longer necks, and hence average neck length, was slightly increased. Once again, the giraffes with longer necks were at an advantage. After many generations, selection for long necks produced the giraffe of today.

Natural selection shapes the characteristics of plants and animals by working on the variation that naturally arises within a group of organisms. Variation is not directed toward advantageous characteristics. The rigors of the environment do not induce variations designed to cope with them. The organisms of a species are individually different, and nature uses the differences to transform the species.

The Origin of Species showed how the simple idea of evolution by natural selection could be used to illuminate a wealth of biological details. Yet some loose ends were left dangling; important questions were left without firm answers. In particular, Darwin had no clear account of the origin and maintenance of variation in natural populations. He assumed that variations would arise and that the capacity for variation in a particular direction would not be diminished by the operation of natural selection; so, for example, as the average neck length of giraffes increases, giraffes with ever longer necks are supposed to appear. Moreover, Darwin's own hazy ideas about inheritance embroiled him in difficulties. He tentatively accepted a theory of "blending inheritance," holding that the characteristics of the progeny result from "mixing" the attributes of the parents. Several of Darwin's early critics pointed out that this theory makes evolution problematical. If an unusual variation arises in a population, then whatever advantages it may confer will be diluted when the distinctive individual mates with other, more mundane, organisms. Quite evidently, the theory of the *Origin* required better answers to questions about variation and inheritance than Darwin was able to supply.

Genes

Ironically, although Darwin recognized the difficulties about variation and inheritance, and though he struggled to overcome them, an important part of the answer was available to him. In 1866 Gregor Mendel published his results about heredity. His article appeared in a relatively obscure journal (the *Proceedings of the Natural History Society of Brünn*), but Mendel sent Darwin a copy. As Darwin labored with questions of heredity in later editions of the *Origin*, Mendel's paper lay on his shelves, apparently unread.

Mendel's great insight was to abandon the theory of "blending inheritance" and to work out an account of "particulate inheritance." He suggested that the characteristics of organisms are governed by *factors*, which are transmitted from one generation to the next and persist unmodified. If we consider a single characteristic of an organism—such as the color of the pods in the pea plants that Mendel investigated—then the form of that characteristic is governed by a particular pair of relevant factors. One of these factors is acquired from each of the organism's parents. When the organism mates it transmits exactly one of the factors to its offspring. On some occasions of mating it may pass on the factor inherited from the mother; on other occasions, the factor acquired from the father will be transmitted.

Moreover, the factors relevant to different characteristics are transmitted independently. A pea plant may pass on the maternal factor for pod color and the paternal factor for length of stem. The particles—the factors—endure, combining in new ways in different progeny.

A simple example from inheritance patterns in humans will illustrate how Mendelian genetics works. My two sons, Andrew and Charles, both have blue eyes. In this they are like my wife and my father, and unlike my mother and me. What explains this pattern of similarities and differences? The Mendelian answer is that there are two relevant factors, a factor for brown eyes and a factor for blue eyes. When a factor for brown eyes appears in an individual, the resulting eye color is (some shade of) brown. Blue-eyed people always have a pair of factors for blue eyes. We can understand the inheritance of eye color in my family as follows. I inherited a factor for brown eyes from my mother, and a factor for blue eyes from my father. As a result, my eyes are brown. My wife has two factors for blue eyes. Hence Andrew and Charles were bound to get at least one factor for blue eyes, since they had to receive one eye color factor from their mother. But, in both cases I transmitted the factor for blue eyes that I acquired from my father. Because of this, both boys have a pair of factors for blue eyes, and their eyes are blue.

This example enables me to introduce some genetic terminology, coined by Mendel's successors, that will prove useful in discussions throughout the book. Mendelian factors are now called *genes*. Alternative forms of a gene are called alleles. For example, the gene for blue eyes and the gene for brown eyes are alleles of the gene for eye color. An individual who has a pair of the *same* alleles for a characteristic is said to be *homozygous* for the characteristic; my wife, my father, and my sons are all homozygous for eye color. Someone who has a pair of *dissimilar* alleles for a characteristic is *heterozygous* for the characteristic; I am heterozygous for eye color. The data given do not determine whether my mother is homozygous or heterozygous for eye color. She might have two alleles for brown eyes, or one allele for brown eyes and one for blue eyes.

In the examples Mendel studied, as well as in the case of blue or brown eye color in humans, the heterozygous individual (the *heterozygote*) has the characteristic manifested by one kind of homozygous individual (*homozygote*). Such cases are usually described by saying that one of the alleles is *dominant*, the other *recessive*. The heterozygote takes on the form of the dominant allele. Thus, in my example, the allele for brown eyes is dominant with respect to the allele for blue eyes. (Alternatively, the allele for blue eyes is recessive with respect to the allele for brown eyes.) Mendel seems to have thought that in every pair of alleles, one would be dominant and the other recessive. We now know that this is wrong. There are many instances in which the attributes of the heterozygote are quite different from those of both homozygotes.

Darwin was not the only person who failed to read Mendel. In fact, Mendel's important ideas languished unappreciated until the turn of the century. Then, in 1900, three scientists, Hugo De Vries, Karl Correns, and Erich von Tschermak, independently rediscovered the principles of Mendelian inheritance. The next decades saw an amazing flurry of discoveries, and Mendel's picture of heredity quickly became both more complete and more complicated.

One relatively trivial change was a change of name. The term *gene* replaced Mendel's *factor*. A more significant step was the explicit recognition of the distinction between the totality of the genes of an organism (the organism's *genotype*) and the characteristics that the organism manifests (its *phenotype*). This distinction was prompted by a growing awareness that the relations between genes (or factors) and characteristics were not as simple as Mendel had believed. Mendel had thought that there was a tight relation between individual factors and individual traits. Each characteristic corresponded to a unique factor, which might exist in alternative forms (alleles) but was represented exactly twice in each organism. After 1900 it quickly became apparent that some characteristics are *polygenic*, that is, they are affected by a number of different genes that are transmitted independently. Many quantitative characteristics of organisms—such as the height of bean plants or the pigmentation of skin in humans and other animals—are polygenic. Moreover, not only can one characteristic in the phenotype be affected by many genes in the genotype, but one gene may also have effects on many different characteristics. Most genes are *pleiotropic*; they contribute to the phenotype in many different ways. So, for example, some genes that affect eye color in the fruit fly *Drosophila melanogaster* also drastically reduce the ability of flies to mate.

Perhaps the most important aspect of the genotype-phenotype distinction is the abandonment of the idea that genes *determine* characteristics. Early Mendelian geneticists tended to focus on simple characteristics which are relatively impervious to the impact of the environment. However, in general, the form of the phenotype is the product not just of the genotype but of a complex interaction between genotype and environment. As seventy years of attempts to isolate exactly what is innate have made clear, it is not easy to abstract from this interaction and to identify the special contribution of the genes.

We are left with the idea that there are hereditary particles, transmitted between generations, that, in conjunction with one another and the environment, produce the phenotypes of organisms. At this point, some obvious questions arise. What are these particles? How are they transmitted? Within three years of the rediscovery of the principles of Mendelian inheritance, these questions had been answered.

Genes and chromosomes

Most organisms are made up of cells. (The exceptions are viruses.) Except in the very simplest cellular organisms (bacteria and some algae) each cell contains a nucleus. The cells of organisms are constantly dividing to form new cells. During the process of cell division, there are important changes within the nucleus. Long threadlike bodies condense, becoming shorter and fatter. We can chart the progress of cell division by observing the arrangement and distribution of these bodies among the daughter cells. The bodies are called *chromosomes*. They obtain their name from the fact that they stain in distinctive ways.

In 1902 two scientists, Walter Sutton and Theodor Boveri, independently proposed that the hereditary factors (genes) are carried on the chromosomes. Their proposal brought Mendelian genetics within the domain of cell biology. In organisms that reproduce sexually, most cell divisions leave the number of chromosomes unaltered. But in *meiosis*, the process in which the sex cells (or *gametes*) are formed, the chromosome number is halved. Of course, when the sperm, the sex cell of the male, unites with the ovum, the sex cell of the female, to produce a fertilized egg (*zygote*), the usual number of chromosomes is reestablished. Here we can glimpse the mechanism of Mendelian inheritance: A zygote obtains exactly half its genes from its male parent because it receives exactly half its chromosomes from the male parent and the chromosomes bear the genes.

A closer look lends further support to the explanation. Most sexually reproducing organisms have chromosomes that pair at meiosis. Each chromosome has a mate, morphologically similar to it, and at meiosis chromosomes line up along with their mates. When the gametes are formed, each gamete receives exactly one chromosome from each pair. When sex cells fuse to form a zygote, not only is the *chromosome number* regained, but the zygote has the *number of matched pairs* that is typical of the species. For example, in humans, each gamete has twenty-three chromosomes, and the zygote has twenty-three pairs.

Armed with some information about cells and meiotic division, we can reconstruct the Mendelian picture of inheritance. Genes are segments of chromosomes. Chromosomes pair just prior to meiotic division; chromosomes that pair with one another are said to be *homologous*. The pairing of homologous chromosomes is, at the same time, a pairing of genes. Each pair of genes affects a particular phenotypic characteristic (or set of characteristics). We can identify a pair of alleles with a distinctive place, or *locus*, on a pair of homologous chromosomes. *Alleles* are alternative pieces of chromosomal material that occur at the same locus.

The new perspective can be illustrated by reverting to my (oversimplified) example about eye color. We envisage the color of a person's eyes as controlled by the material present at a particular locus on a particular pair of human chromosomes. If a certain type of chromosomal material (the allele for brown eyes) is found at that locus on either or both of the chromosomes, then the person has brown eyes. If a different type of chromosomal material (the allele for blue eyes) is present at the locus on *both* of the chromosomes, then the person has blue eyes. It is not difficult to retell the story I told about my family, charting the transmission of the chromosomes.

The chromosomal picture suggests a new complication of Mendel's account of inheritance. Mendel believed that genes (factors) are transmitted independently. If I receive a certain allele from my mother, then that is not supposed to affect the likelihood of my receiving any other allele from my mother. In our new terminology, this is tantamount to saying that all loci are independent. Now it is fairly clear that this will not be so. Chromosomes contain a large number of loci—there are many more genes than chromosomes. The alleles occupying all the loci on the same chromosome will tend to be transmitted together. (See below for further details.) Genes that occur on nonhomologous chromosomes (chromosomes that do not pair with one another) may be transmitted independently. Genes whose loci are on the same chromosome pair will not be independent.

Before I add further wrinkles to the story, let us take stock. We may seem to have wandered a long way from the theory of evolution and to have pursued irrelevant details about heredity. However, we have now reached a point at which it is possible to address one of the questions that vexed Darwin. The theory of evolution studies the ways in which nature exploits the *genetic variation* in a population of organisms, that part of the variation among individuals stemming from differences in genotypes. But how does genetic variation arise? To answer this question, we have to understand how there can be alternative alleles at a locus. A simple proposal is that, prior to the end of meiosis, the chromosomal material at a locus may be altered, producing a *mutant allele*. In the early decades of this century, long before the constitution of the chromosomes was known, this simple proposal won the acceptance of geneticists. Thanks to the labors of a number of scientists—most notably H. J. Muller—a great deal was learned about the process *(mutation)* in which mutant alleles are formed. Finally, as we shall see in the next section, molecular genetics has provided a clear and detailed account of what happens in this process.

Mutation is the ultimate source of the genetic variation in a population. But it is important to recognize that, even in the absence of mutations, new genotypes are constantly being formed, yielding new material on which natural selection can act. I received half of my genes from my father and half from my mother, but the combination is probably unique in human history. Now it may appear that there is a limit to the possibilities of assortment. Because genes are linked (whole chromosomes apparently being transmitted together), the number of biologically possible genotypes seems to be substantially less than the number of mathematically possible combinations of alleles.

It is time to add another detail to the chromosomal picture. Although genes that occur on the same chromosome (more exactly, relatively close on the same chromosome) are linked, the linkage is not usually complete. The progress of meiosis does not merely bring homologous chromosomes alongside one another. Homologous chromosomes may also break and rejoin in such a way that part of one chromosome becomes attached to the complementary part of its mate. Figure 1.1 will make this idea clear. Here, A, B, A', B' are chromosomal segments, which *cross over* (middle picture) to form new chromosomes in the gametes. The new chromosomes are *recombinant chromosomes*, and the entire process is called *recombination*. Recombination is quite likely to occur between loci that are some distance apart. It is relatively improbable—though not impossible—between loci that are close together. While the phenomenon of linkage shows that Mendel was wrong to suppose that genes are transmitted independently, the fact that linkage is incomplete allows for the constant production of novel chromosomes (and thus novel sequences of genes) through recombination.

Cell biology and genetics have had an enormously fruitful marriage. Since 1910, when the great American biologist T. H. Morgan assembled a remarkable group of talented students and coworkers in the famous "flyroom" at Columbia University, the chromosomal perspective on the arrangement and transmission of genes has yielded enormous

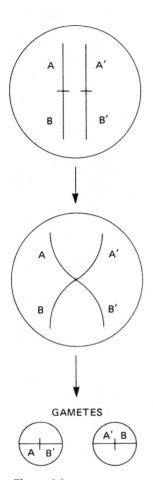

Figure 1.1
Crossing over and recombination.

insights into the mechanisms of inheritance. I shall not try to summarize all of the important discoveries that have been made. Instead, I shall skip several decades and consider another major breakthrough.

Genes and molecules

Once genes had been identified as segments of chromosomes, it was natural for biologists to hope that an understanding of the chemical structure of the chromosomes might explain some features of gene action and interaction. The hope was fulfilled in a sequence of brilliant achievements. In the 1940s and early 1950s it became clear that the genetic material in cellular organisms is DNA (deoxyribonucleic acid). (The crucial results were obtained by Avery, MacLeod, and McCarty in 1944 and Hershey and Chase in 1952.) These discoveries set the stage for an intense struggle to find the chemical structure of DNA. The struggle ended in 1953 when James Watson and Francis Crick announced a model for DNA: the famous double helix.

DNA molecules consist of a pair of sugar phosphate chains that spiral around one another. Projecting inside each chain, like the rungs of a ladder, are *bases*, or *nucleotides*, which match up with one another in specific ways. Discovery of the structure of DNA paved the way for an understanding of how the genes work to produce (with the aid of the environment) phenotypic characteristics. Before 1953 scientists had recognized that the primary function of many genes seems to be to produce molecules that are needed for chemical reactions within the cell. Studies by Beadle, Tatum, and others suggested that genes direct the formation of *proteins*. Proteins are complicated molecules, built up out of long chains of *amino acids*. (The chains are called *polypeptides*.) We now know that many genes direct the production of polypeptides. The polypeptides then join together to form proteins. The proteins then participate in a sequence of chemical reactions. At the end of all this hectic chemical activity stand observable effects like eye color.

The genes direct the formation of particular polypeptides because the genetic material, DNA, carries specific information. The structure of the polypeptides formed depends on the sequence of nucleotides that is present within the gene. The order of the bases carries a message. If that order is changed, the message is changed, and a different polypeptide can result.

This fact is the key to understanding mutation. A mutation involves a change in the sequence of bases on a strand of DNA. New nucleotides may be inserted; old nucleotides may be deleted; one nucleotide may be replaced by another. When any of these events occur, the resultant molecule—the mutant gene—will, like all segments of DNA, retain the capacity for replicating itself. But modification of the sequence of bases is likely to lead to the production of a different chain of amino acids. If a deviant amino acid chain is formed, then the protein of which it is a part may be unable to engage in the same chemical reactions as the ordinary protein. Thus the chemical economy may be affected, either slightly or drastically. If the effects are sufficiently large, then there will be an observable change in phenotype.

In discussing the structure and function of DNA, I have claimed that *many* genes direct the production of polypeptide chains. There is an important distinction between two kinds of genes. *Structural* genes direct the formation of chains of amino acids. Other genes, *regulatory* genes, act like switches, turning the protein makers on and off as needs require. The mechanics of gene regulation in some simple organisms (like bacteria) is relatively well understood, but the analogous processes in higher organisms are currently the subject of intensive research. However, a number of studies in the last decades have made it clear that a significant portion of genetic material consists of segments of DNA that organize the structural genes.

Once again, it may seem that these investigations have little bearing on the theory of evolution. But molecular biology will prove relevant to our discussion in two different ways. The molecular perspective on gene mutation and gene functioning enables us to see what kinds of genetic variations may arise. In addition, some techniques of recent molecular biology have enabled scientists to compare the molecular structures of the proteins in related species. It is now possible to examine the differences between kindred proteins in humans and chimpanzees (to cite just one example), and, because the proteins reflect the structure of the DNA, this helps to expose the molecular steps that have taken place in the transformation of species.

Genes and populations

With some understanding of the main ideas of genetics, we can now approach contemporary evolutionary theory, the *synthetic theory of evolution,* or the *neo-Darwinian synthesis* as it is sometimes called. The synthesis joins the ideas of genetics with Darwin's theory of evolution. In the 1930s R. A. Fisher, J. B. S. Haldane, and Sewall Wright independently laid the foundations of the mathematical theory of population genetics. Building on their achievements, Theodosius

Dobzhansky, Ernst Mayr, G. G. Simpson, and Julian Huxley produced the "modern synthesis."

A central concept in Darwin's thinking, which becomes preeminent in the writings of his twentieth-century successors, is the concept of a *population*. From the viewpoint of evolutionary theory, a *population* is a group of organisms that freely interbreed in nature. As a result of the process of interbreeding, the genotypes of individuals are constantly being broken up and the constituent genes are brought together in new ways. Using an obvious metaphor, we may speak of the population as having a common *gene pool*. From the perspective of genetics, evolutionary changes consist in alterations of the gene pool of a population: Some genes obtain greater representation in the pool; others become less frequent or drop out altogether.

The mathematical part of population genetics studies the ways in which the frequencies of genotypes in a population change when the organisms comprising a population are subjected to specific pressures. Its most elementary results concern the evolution of very large populations in which no biases in mating are assumed to occur. Using the ideas of genetics, it is relatively easy to work out what will be expected to happen in this kind of population, given that (1) two alleles are present at a loc s with a specified relative frequency, (2) the allele frequencies are not disturbed by mutation or migration from one population to another, and (3) no combination of alleles confers any advantage on its possessor. Under these circumstances, the population reaches *evolutionary equilibrium*: The relative frequency of each allele remains constant. This result was discovered independently in 1908 by G. H. Hardy and W. Weinberg. It is usually referred to as the *Hardy-Weinberg law*.

Evolutionary equilibrium can easily be disturbed, and mathematical population genetics studies in detail what will follow from various kinds of disturbance. We can allow for effects of mutation and migration of alleles. Moreover, we can consider how genotype frequencies will change when some allelic combinations are taken to be more advantageous than others.

I shall illustrate the elementary ideas of population genetics with an extremely simple example. Let us suppose that we have two alleles, A and a, at a locus, in a very large population of organisms. Assume that initially all the individuals in the population are heterozygous; that is, they have the genotype Aa. Further, let us suppose that the possible genotypes are not equally good: Aa individuals always survive from birth to maturity, AA individuals survive to maturity half the time, and aa individuals never survive to maturity. What will be the distribution of genotypes among the mature individuals of the next generation? We can work this out as follows. Two types of gametes, carrying the genes *A* and *a*, respectively, will be formed in equal numbers. Assuming random mating, four types of zygote formation will occur in equal numbers: $A \times A, A \times a, a \times A, a \times a$. So zygotes with the genotypes *AA, Aa, aa* will be formed in the ratio 1 : 2 : 1. All of the *aa* zygotes will fail to mature, and half of the *AA* zygotes will do likewise. Hence, the mature organisms of the next generation will consist of *AA* individuals and *Aa* individuals in the ratio 1 : 4.

I introduce this simple artificial example to show how the Mendelian and Darwinian ideas may be combined so that, with a little mathematics, it is possible to work out the genetical development of a population. Theoretical population genetics is able to cope with much more realistic examples (although as the treatment becomes more applicable to real organisms, the mathematics becomes much more difficult). It can provide estimates of the number of generations required for an initially scarce allele to achieve the status of the only allele at its locus, that is, for the allele to become *fixed* in the population. It can show that, under certain conditions, there will be an equilibrium value for the relative frequency of two alleles. Such results are directly relevant to the program of evolutionary theory. For they help us to see the general characteristics of the alterations in the gene pool that are the basis of evolutionary change.

Darwin's fundamental idea that there is a struggle for existence, and that some organisms are better adapted to succeed in surviving and reproducing enters the mathematical theory of population genetics in an obvious way. We may speak of the *fitness* (or *selective value*) of an allelic combination, or of a whole genotype, identifying the extent to which organisms with that allelic combination (genotype) are likely to be successful in the competition for survival and reproduction. Mathematical population genetics offers precise theorems about changes in the frequency of genotypes (or allelic combinations), given numerical specifications of the relevant fitness (selective values).

In the light of the earlier discussions of genetics, it should be clear that this conception of fitness and the theory that articulates it are both idealizations. The prospects for the bearer of a particular allelic combination are not likely to be a function of that combination alone. Instead, the chances for survival and successful reproduction will almost always depend on other features: the external environment, the character of the population to which the organism belongs, the other genes that make up the genotype. Genes act in concert. A gene that is helpful against one genetic background may be deleterious against another.

Moreover, the phenotype and its promotion of reproductive success are both functions of the environment. A gene producing pigmentation that facilitates camouflage in one habitat may be responsible for unwelcome visibility under different circumstances. For almost all pairs of alleles, we cannot specify some unique number that represents the *absolute* fitness.

Nevertheless, it is sometimes possible to abstract from the vicissitudes of genetic background, population characters, and environment. Just as physicists can sometimes make idealizations, treating real surfaces as if they were frictionless planes, so evolutionary biologists are frequently able to compare the selective values of different allelic pairs. Given what is known about the likely genetic background, the population, and the environment, it may be clear that particular combinations of alleles at a locus have specifiable effects; one may see, for example, that a certain pair of alleles drastically reduces the chances of survival. So, for all practical purposes, it may be possible to assign selective values to allelic pairs. However, the success of this practice should not lead us to forget the general moral of the last paragraph. When we can specify values of fitness, that is because it is possible to hold constant the variables on which fitness generally depends.

With this cursory look at the concept of selective value, we are ready to integrate Darwin's idea of natural selection with the ideas of genetics. Evolutionary change is change in the genetic constitution of a population. This type of change can come about as the result of a number of factors. Immigration and emigration of organisms will bring new alleles and new allelic combinations into the population and will cause others to disappear. Mutation will lead to the formation of new alleles. The major claim of a *Darwinian* theory of evolution is that the *principal* factor of change is natural selection: *The most important evolutionary changes come about because some allelic pairs are fitter than others, and these obtain greater representation for their constituent alleles in subsequent generations.* To take a simple example, if the *AA* homozygote confers significant advantages, then the *A* allele is likely to increase in the population. (It may even become fixed.) Mathematical population genetics tells us *precisely* how the forces of immigration, emigration, mutation, and natural selection produce evolutionary change.

Mathematical population genetics is not the only part of the study of the genetics of natural populations that is relevant to evolutionary theory. Once Darwin's ideas had been recast in terms of changes in the frequency of alleles, important empirical investigations could begin. One significant achievement, the result of years of work by Dobzhansky and his associates, has been a more detailed understanding of the variation that is naturally present in a population. Darwin supposed that new variations in a population will constantly arise, and that this variation can be maintained, so as to provide material for the action of natural selection. These suppositions provoke obvious questions. How do variations arise? Do they occur with sufficient frequency for significant evolutionary change? How is the variation of a natural population preserved?

We have already encountered the answer to the first question. Variation is produced in two ways: through recombination of existing alleles and through the introduction of new alleles by mutation. Of these, the second is the more fundamental. Through experimental study of mutation rates in a variety of organisms (but most extensively in flies of the genus *Drosophila*), Dobzhansky and other scientists have shown that, while the rates of mutation at individual loci are low, a considerable number of mutations arises in each generation of a typical natural population.

The final question is more difficult. It might seem that there will be a typical evolutionary scenario: A population is likely to consist of homozygotes for the most favorable allele; but, sooner or later, some lucky mutant arises with higher selective value and spreads itself through the population, so that selection produces a new group of (different) homozygotes. However, Dobzhansky's detailed studies revealed that evolution does not work this way. At many loci, alternative alleles are maintained in the population.

How is this achieved? One way for genetic variation to be maintained is for the heterozygote to have higher fitness than either homozygote. A classic—although tragic—example of this phenomenon is the persistence of sickle-cell anemia. Those who suffer from sickle-cell anemia are homozygous for a recessive gene that directs the formation of abnormal hemoglobin molecules. The abnormal hemoglobin is functionally deficient, and the disease usually leads to an early death. However, heterozygous individuals produce enough normal hemoglobin to thrive, and they reap an extra dividend. Their genotype confers protection against some forms of malaria. So, in populations that are subject to these types of malaria, selection maintains the allele for the deviant hemoglobin (albeit at low frequency).

Another way for variation to be maintained by selection is for the fitness of an allele to increase as it becomes less frequent in the population. Studies of flies of the genus *Drosophila* have revealed an interesting result. Males with certain gene arrangements are more successful in mating when those arrangements are rare. As their success spreads the gene arrangement through the population, the advantage of having it decreases, so that, when it becomes more prevalent, another arrangement is favored. Natural selection maintains a balance.

Dobzhansky attempted to defend two main conclusions. He argued that there is a great deal of genetic variation in natural populations, a result that was further confirmed in the 1960s when molecular biology provided techniques for measuring allelic variation. He also contended that the variation was maintained by natural selection— by means such as the superior fitness of heterozygotes or the increased fitness of rare genetic combinations. The second claim is more controversial. Although biologists agree that natural selection can preserve variation in some cases (like those that I have mentioned), some population geneticists argue (as will be seen) that much genetic variation arises as the result of chance events.

The origin of diversity

We are now ready to take a closer look at the main evolutionary thesis, the claim that one species of organism can give rise to radically different forms. To make this claim precise, it is necessary to overcome a difficulty that I glossed over in presenting Darwin's original views: What exactly is a species? Darwin had trouble with this question. The definition current in his day identified species as collections of organisms whose descendants necessarily belong to the same species as their ancestors. Hence to speak of *species* as evolving appeared self-contradictory. In presenting his evolutionary thesis, Darwin oscillated between saying that species really do evolve and denying that species (that is, species in the traditional sense) exist at all.

A clear theory of evolution needs a new concept of species. The "modern synthesis" brought the required new concept, the *biological species concept*, articulated and defended by Ernst Mayr. I shall consider the concept only is it applies to organisms that reproduce sexually. For such organisms, we can define the idea of *reproductive isolation* as follows. Two populations of organisms that overlap geographically are said to be *reproductively isolated* from one another if there is no significant gene flow between the populations. Although organisms from one population occur in the same region as organisms from the other population, they do not breed in the course of nature (or, at least, they do not do so very frequently). Different species may be very similar indeed in their morphological characteristics. A classic example is a cluster of six species of mosquitoes (belonging to the genus *Anopheles*). Although they are extremely difficult to tell apart, the six species were distinguished by the criterion of reproductive isolation. Interestingly, this distinction solved an important problem about the transmission of malaria. Before the mosquito species were distinguished, it seemed impossible to understand why malarial infections were frequent in some areas and rare or nonexistent in others. Once the distinction was made, the carriers were recognized, and the facts fell into place.

Reproductive isolation signals the presence of *isolating mechanisms* between the populations. Isolating mechanisms can take a number of different forms, including mechanical difficulties of mating, absence of sexual attraction, inviability or sterility of hybrids, and other factors as well. Once we have some understanding of the nature of isolating mechanisms and of the genetic, anatomical, physiological, and behavioral characteristics with which they are associated, then it is possible to ask whether two populations that do not occur in the same region at the same time constitute distinct species. If there are isolating mechanisms between them, then, even if they had been in the right place at the right time, there would still have been no significant gene flow. Thus the populations would belong to different species. On the other hand, if no isolating mechanisms are present, then interbreeding is debarred only by the quirks of location in space and time. The two populations are spatiotemporally separated populations of the same species.

Dividing species by the criterion of reproductive isolation makes the origin of species a crucial step in the development of diversity. When two populations become reproductively isolated, they go their separate ways as genetic units. Changes in the genetic constitution of one population are not automatically reflected in the genetic constitution of the other population. Thus, once reproductive isolation has been achieved, the populations are likely to diverge further and further. Conversely, if reproductive isolation is not achieved, geographical separation of two populations may only amount to a temporary protection of the genetic differences between the populations. If the geographical barrier is surmounted and the populations are reunited, gene flow between the populations will occur freely. This means that there will be effectively one population, which will respond as a single interbreeding group to the demands of the environment.

How then do new species come to be? (How does *speciation* occur?) Ernst Mayr has argued cogently that a crucial first step to *reproductive* isolation is *geographical* isolation. He envisages a small subgroup of a spatially extended population becoming cut off from the main body of the population. Within this subgroup, perhaps because of distinctive environmental pressures, there are important genetic changes, which eventually make the descendants of the separated subgroup reproductively isolated from their contemporaries who have developed from the main population. After the breakdown of the geographical barrier, the new species comes into contact with the old species, and the isolating mechanisms protect the integrity of the gene pool of each species. (Some biologists have proposed that selection against unfit hybrids may even serve to accentuate isolating mechanisms, once contact has been reestablished.)

Mayr's model explains how species may multiply by the splitting off of new species from old. Many neo-Darwinians would also stress the possibility that a population that does not divide to yield new species may become radically transformed over the course of a number of generations. This process, too, can be viewed as the birth of a new species. For, if we look at a *lineage*, a sequence of populations of organisms related by descent, we may find that the genetic constitution of successive populations is so modified that populations occurring late in the sequence are reproductively isolated from those present at the early stages. The populations did not actually have the chance to breed together—but if they had, free interbreeding would not have occurred.

The origin of new species is a crucial step in the diversification of life. But it is only the first step. Taxonomists group organisms in ever more inclusive collections. Kindred species are collected into the same *genus*, kindred *genera* into the same *family*, kindred *families* into the same *order*, kindred *orders* into the same *class*, kindred *classes* into the same *phylum*, kindred *phyla* into the same *kingdom*. (The species *Canis lupus*, the wolf, belongs to the genus *Canis*, encompassing dogs, wolves, dingos, and coyotes; this genus belongs to the family Canidae, which comprises the foxes and jackals as well; the family belongs to the order of carnivores, an order within the class of mammals; this class in turn belongs to the phylum of chordates, that part of the kingdom of animals containing animals with some internal supportive structure.)

How are we to understand the emergence of new families, new classes, or new phyla? The answer builds on the account of speciation. A new family, class, or phylum first emerges as a new species (possibly a small number of new species). In the course of acquiring reproductive isolation from its ancestral population, it has also gained some distinctive feature that makes it possible for it to invade a new *adaptive zone*. (The idea of an adaptive zone was suggested by the great American biologist G. G. Simpson. Simpson's main contributions to the modern synthesis lie in his detailed investigations in *paleontology*—the study of the course of evolution in the past.) Roughly, the adaptive zone of a group of organisms represents the way in which that group uses the environment to make a living.

Evolutionary theorists propose that, in the history of life, new species have sometimes been able to enter previously unoccupied adaptive zones. The first amphibians and the early birds found original ways to tap the resources of nature. Once a new adaptive zone has been entered, then the way is open for descendants of the pioneering species to diversify extensively, producing so wide a range of forms that they become recognized as making up a new genus, a new family, a new order, a new class, a new phylum, even a new kingdom. Of course, the extent to which diversification proceeds depends on the breadth of the adaptive zone, the range of alternative possibilities that it allows. For our purposes, however, what is important is the nature of the first step. A new species, possibly quite similar to its ancestors, gains some distinctive characteristic that enables it to invade a new adaptive zone. Initially, its hold on the zone may be precarious; all the ancestral traits may interfere with exploiting the new resources. Retrospectively, the pioneering species, which may seem primitive and clumsy in comparison with its flourishing descendants, appears as the first member of a new genus (family, order, class, phylum, or kingdom).

Small steps toward perfection?

At this point I want to consider two further ideas of evolutionary theory that are important both because they are matters of current debate and because they give rise to misunderstandings on which the enemies of evolution can capitalize. Darwin championed the thesis of *evolutionary gradualism*. He believed that evolution is a slow and gradual process that takes place by small, discrete steps. Contemporary neo-Darwinians have followed Darwin's lead. They use the ideas of genetics to give substance to the gradualist thesis. The orthodox view is that many individual genetic changes are needed for a population to achieve reproductive isolation from its ancestors. A large number of new alleles must enter the gene pool and become prevalent in it. Speciation is gradual, a process in which small genetic changes accumulate until a new species is formed.

There are many well-understood cases of genetic changes in natural populations. Within the last 150 years, several species of moths (most notably the peppered moth, *Biston betularia*) have shown an increasing prevalence of *melanic* forms. (The melanic individuals are black, and have an advantage in being less conspicuous to predators in areas where industrial pollution has blackened tree trunks.) Even more striking have been the numerous cases in which insects have developed resistance to pesticides. Careful investigation has traced such changes to a small number of allelic substitutions. Orthodox neo-Darwinians believe that these modifications are the stuff of which large-scale evolution is made. The production of new species, new genera, and so forth simply consists of more of the same.

I want to emphasize that this is a proposal about the "tempo and mode" of evolution that is *independent* of the claims surveyed in previous sections. No inconsistency threatens an evolutionary theorist who accepts those other claims but who denies that species are born through a slow sequence of small genetic changes. As we shall see in chapter 5, there are heterodox evolutionary theorists who challenge the standard gradualist view. Creationists try to adapt these challenges to serve the interests of their own cause.

A second important and controversial topic concerns the extent to which evolution gives rise to *adaptations*. The fundamental idea of natural selection is that some alleles confer advantages on their bearers, and that these are likely to spread through the gene pool of the population. Does this mean that all the characteristics produced in the evolutionary process are *adaptations*, attributes that represent an optimal solution to the problems that the environment poses for the organism? No. It was obvious to Darwin, and it has continued to be obvious to all of his successors, that not every characteristic that is produced in the evolutionary modification of a population is adaptively advantageous to the organisms that possess it. Darwin referred, a bit vaguely, to (unknown) principles of "correlation and balance," suggesting that a feature of an organism may come to be present because it is connected with another characteristic that bestows a competitive advantage. His contemporary successors can be more specific. Once it is recognized that most genes are pleiotropic (that is, they participate in forming many phenotypic traits), that genes are linked, that genes work together, and so forth, the emergence of nonadaptive characteristics seems inevitable. An allele may be selected because it helps to secure important advantages, which outweigh some neutral, or even slightly counterproductive, effects that it brings in its train.

There are other reasons for thinking that nonadaptive characteristics can become prevalent in a population. Since the 1930s, it has been apparent that chance events, the accidents of mating, for example, will play a role in evolution. The term "genetic drift" has been coined to cover processes in which gene frequencies are changed as the result of chance. No contemporary evolutionary biologist doubts that genetic drift exists. The sorting out of genotypes in successive populations is

analogous to tossing a large number of coins: The "theoretically expected value" (half heads in the case of the coins) is likely to be slightly
different from what we actually find. (It would be remarkable to get
exactly 500 heads in 1,000 coin tosses.) But the significance of drift is
controversial. *Neutralists* maintain that the role of natural selection in
evolution has been overemphasized. They hold that most genetic variation results from the interplay between mutation and drift. Alternative
allelic combinations are supposed not to differ in fitness. *Selectionists*
respond by arguing that genetic variation is preserved by natural
selection, which maintains more than one allele by a balance of selective
forces (as, for example, in the case of sickle-cell anemia). The debate
concerns details of the workings of evolution. It is predicated on the
assumption that evolution occurs.

Evolutionary theory is not a closed subject in which all the interesting
work has been done. There are still important issues to be resolved
about the mechanics of evolution. Yet I hope that my brief survey of
two major areas of controversy—the debates about gradualism and
the prevalence of adaptations—reveals how the disputes do not threaten
major evolutionary ideas. Though evolutionary biologists argue
vigorously, the foundations of their discipline are not tottering.

The history of life

No self-respecting introduction to the main ideas of evolutionary theory
can conclude without some mention, however sketchy, of the story
of life. So I shall conclude this chapter with a brief survey of organic
history, from the algae to us. My aim is to mention a few important
incidents that can provide a context for later discussions. I am also
out to set a new record: 4 billion years of history in under three pages.

The earth was formed about 4.5 billion years ago. Biologically significant molecules appeared relatively quickly. From a primitive atmosphere, rich in simple molecules (hydrogen, nitrogen, water,
methane, ammonia), organic compounds were generated. Eventually,
self-replicating molecules—like DNA—appeared, along with amino
acids and proteins. The first simple organisms, probably unicellular
organisms without a nucleus, were made from these basic constituents.
A major evolutionary step was taken when the earliest algae, the blue-
green algae "invented" photosynthesis. This achievement enabled the
blue-green algae to dominate life on earth for about half of our planet's
history. Only the emergence of the green algae, the first organisms
with a cellular nucleus, about 1.5 billion years ago, challenged the
supremacy of their blue-green predecessors.

When we reach the last billion years, the pace of change becomes more rapid. About 600 million years ago, there was a sudden proliferation of marine invertebrates. Sponges, worms, mollusks, and many similar creatures occur in great profusion in the so-called Cambrian explosion. About 100 million years later, the first vertebrates arose. These were jawless fishes, somewhat like lampreys. The next 100 million years saw the evolution of many kinds of fish. One line of descent gave rise to the amphibians, which, about 400 million years ago, became the first vertebrates to invade land.

The amphibians were quickly succeeded by the reptiles. (What's 100 million years in geological time?) Roughly 300 million years ago, the earliest reptiles developed a method of reproduction (the *amniote egg*) that enabled them not only to live on land (as the amphibians had done) but also to breed on land. By the end of the *Paleozoic* era (about 230 million years ago) the reptiles had become the ruling land animals.

The next era, the *Mesozoic*, saw the rise and fall of the most famous extinct group of animals: the dinosaurs. Apart from their other accomplishments, these sometimes spectacular beasts deserve to be mentioned because of their ability to fascinate children—some of whom grow up to be paleontologists. But the dinosaurs were also remarkably successful. From quite early in the Mesozoic until their mysterious extinction about 70 million years ago, they dominated our planet. Beside them, two other Mesozoic developments seem almost insignificant. About 180 million years ago, different groups of reptiles gave rise to the mammals and the birds. The first mammals were small, nocturnal, insectivorous creatures that seem merely to have taken advantage of whatever space the dinosaurs left vacant. Nothing much indicated that they had a bright future.

In the most recent era, the *Cenozoic* (which began about 65 million years ago), the mammals and birds came into their own. With the dinosaurs gone, the birds and mammals have diversified extensively. Within the class of mammals, the species that are most familiar to us have appeared. Primitive carnivores have given rise to dogs and bears, hyenas and cats, seals and sea lions. Primitive herbivores have given rise to cows and sheep, deer and antelope, horses and rhinoceroses. Another evolutionary development, more significant to us, but no more dramatic than many other evolutionary events, is the diversification of the *Primates*. The first Primates were primitive insectivorous mammals of the late Mesozoic. The end, we may feel, belies the unspectacular beginning. One line of descent leads to the great apes— and their close cousin, *Homo sapiens*.

While this sequence of major evolutionary changes has occurred, there have been innumerable other events in which species have been born, become modified, and become extinct. Moreover, throughout the entire process the environmental conditions have constantly changed. Climatic conditions have altered dramatically. Mountain ranges have been built. The continents have separated from an original "supercontinent." Old land connections have been broken; new connections have been forged. The scope of the drama is immense. Darwin summed up his own feelings of awe in the concluding sentence of *The Origin of Species*: "There is grandeur in this view of life, with its several powers, having been originally breathed into a few forms or into one; and that, whilst this planet has gone cycling on according to the fixed law of gravity, from so simple a beginning endless forms most beautiful and most wonderful have been, and are being, evolved" (Darwin 1859/ Mayr 1964, 490).

2

Believing Where We Cannot Prove

Opening moves

Simple distinctions come all too easily. Frequently we open the way for later puzzlement by restricting the options we take to be available. So, for example, in contrasting science and religion, we often operate with a simple pair of categories. On one side there is science, proof, and certainty; on the other, religion, conjecture, and faith.

The opening lines of Tennyson's *In Memoriam* offer an eloquent statement of the contrast:

Strong Son of God, immortal love,
Whom we, that have not seen Thy face,
By faith, and faith alone, embrace,
Believing where we cannot prove.

A principal theme of Tennyson's great poem is his struggle to maintain faith in the face of what seems to be powerful scientific evidence. Tennyson had read a popular work by Robert Chambers, *Vestiges of the Natural History of Creation,* and he was greatly troubled by the account of the course of life on earth that the book contains. *In Memoriam* reveals a man trying to believe where he cannot prove, a man haunted by the thought that the proofs may be against him.

Like Tennyson, contemporary Creationists accept the traditional contrast between science and religion. But where Tennyson agonized, they attack. While they are less eloquent, they are supremely confident of their own solution. They open their onslaught on evolutionary theory by denying that it is a science. In *The Troubled Waters of Evolution,* Henry Morris characterizes evolutionary theory as maintaining that large amounts of time are required for evolution to produce "new kinds." As a result, we should not expect to see such "new kinds" emerging. Morris comments, "Creationists in turn insist that this belief is not scientific evidence but only a statement of faith. The evolutionist seems to be saying, Of course, we cannot really *prove* evolution, since this requires ages of time, and so, therefore, you should accept it as a proved fact of science! Creationists regard this as an odd type of logic, which would be entirely unacceptable in any other field of science" (Morris 1974b, 16). David Watson makes a similar point in comparing Darwin with Galileo: "So here is the difference between Darwin and Galileo: Galileo set a demonstrable *fact* against a few words of Bible poetry which the Church at that time had understood in an obviously naive way; Darwin set an unprovable *theory* against eleven chapters of straightforward history which cannot be reinterpreted in any satisfactory way" (Watson 1976, 46).

The idea that evolution is conjecture, faith, or "philosophy" pervades Creationist writings (Morris 1974a, 4–8; Morris 1974b, 22, 172; Wysong 1976, 43–45; Gish 1979, 11–13, 26, 186; Wilder-Smith 1981, 7–8). It is absolutely crucial to their case for equal time for "scientific" Creationism. This ploy has succeeded in winning important adherents to the Creationist cause. As he prepared to defend Arkansas law 590, Attorney General Steven Clark echoed the Creationist judgment. "Evolution," he said, "is just a theory." Similar words have been heard in Congress. William Dannemeyer, a congressman from California, introduced a bill to limit funding to the Smithsonian with the following words: "If the theory of evolution is just that—a theory—and if that theory can be regarded as a religion . . . then it occurs to this Member that other Members might prefer it not to be given exclusive or top billing in our Nation's most famous museum but equal billing or perhaps no billing at all."

In their attempt to show that evolution is not science, Creationists receive help from the least likely sources. Great scientists sometimes claim that certain facts about the past evolution of organisms are "demonstrated" or "indubitable" (Simpson 1953, 70, 371; also Mayr 1976, 9). But Creationists also can (and do) quote scientists who characterize evolution as "dogma" and contend that there is no conclusive proof of evolutionary theory (Matthews 1971, xi; Birch and Ehrlich 1967, 349; quoted in Gish 1979, 15–16; similar passages are quoted in Morris 1974a, 6–8, and in Wysong 1976, 44). Evolution is not part of science because, as evolutionary biologists themselves concede, science demands proof, and, as other biologists point out, proof of evolution is not forthcoming.

The rest of the Creationist argument flows easily. We educate our children in evolutionary theory as if it were a proven fact. We subscribe officially, in our school system, to one faith—an atheistic, materialistic faith—ignoring rival beliefs. Antireligious educators deform the minds of children, warping them to accept as gospel a doctrine that has no more scientific support than the true Gospel. The very least that should be done is to allow for both alternatives to be presented.

We should reject the Creationists' gambit. Eminent scientists notwithstanding, science is not a body of demonstrated truths. Virtually all of science is an exercise in believing where we cannot prove. Yet, scientific conclusions are not embraced by faith alone. Tennyson's dichotomy was too simple.

Inconclusive evidence

Sometimes we seem to have conclusive reasons for accepting a statement as true. It is hard to doubt that $2 + 2 = 4$. If, unlike Lord Kelvin's ideal mathematician, we do not find it obvious that

$$\int_{-\infty}^{+\infty} e^{-x^2} \, dx = \sqrt{\pi},$$

at least the elementary parts of mathematics appear to command our agreement. The direct evidence of our senses seems equally compelling. If I see the pen with which I am writing, holding it firmly in my unclouded view, how can I doubt that it exists? The talented mathematician who has proved a theorem and the keen-eyed witness of an episode furnish our ideals of certainty in knowledge. What they tell us can be engraved in stone, for there is no cause for worry that it will need to be modified.

Yet, in another mood, one that seems "deeper" or more "philosophical," skeptical doubts begin to creep in. Is there really anything of which we are so certain that later evidence could not give us reason to change our minds? Even when we think about mathematical proof, can we not imagine that new discoveries may cast doubt on the cogency of our reasoning? (The history of mathematics reveals that sometimes what seems for all the world like a proof may have a false conclusion.) Is it not possible that the most careful observer may have missed something? Or that the witness brought preconceptions to the observation that subtly biased what was reported? Are we not *always* fallible?

I am mildly sympathetic to the skeptic's worries. Complete certainty is best seen as an ideal toward which we strive and that is rarely, if ever, attained. Conclusive evidence always eludes us. Yet even if we ignore skeptical complaints and imagine that we are sometimes lucky enough to have conclusive reasons for accepting a claim as true, we should not include scientific reasoning among our paradigms of proof. Fallibility is the hallmark of science.

This point should not be so surprising. The trouble is that we frequently forget it in discussing contemporary science. When we turn to the history of science, however, our fallibility stares us in the face. The history of the natural sciences is strewn with the corpses of intricately organized theories, each of which had, in its day, considerable evidence in its favor. When we look at the confident defenders of those theories we should see anticipations of ourselves. The eighteenth-century scientists who believed that heat is a "subtle fluid," the atomic theorists who maintained that water molecules are compounded out of one atom of hydrogen and one of oxygen, the biochemists who identified protein as the genetic material, and the geologists who thought that continents cannot move were neither unintelligent nor ill informed. Given the evidence available to them, they were eminently reasonable in drawing their conclusions. History proved them wrong. It did not show that they were unjustified.

Why is science fallible? Scientific investigation aims to disclose the general principles that govern the workings of the universe. These principles are not intended merely to summarize what some select groups of humans have witnessed. Natural science is not just natural history. It is vastly more ambitious. Science offers us laws that are supposed to hold universally, and it advances claims about things that are beyond our power to observe. The nuclear physicist who sets down the law governing a particular type of radioactive decay is attempting to state a truth that holds throughout the entire cosmos and also to describe the behavior of things that we cannot even see. Yet, of necessity, the physicist's ultimate evidence is highly restricted. Like the rest of us, scientists are confined to a relatively small region of space and time and equipped with limited and imperfect senses.

How is science possible at all? How are we able to have any confidence about the distant regions of the cosmos and the invisible realm that lies behind the surfaces of ordinary things? The answer is complicated. Natural science follows intricate and ingenious procedures for fathoming the secrets of the universe. Scientists devise ways of obtaining especially revealing evidence. They single out some of the things we are able to see as crucial clues to the way that nature works. These clues are used to answer questions that cannot be addressed by direct observation. Scientific theories, even those that are most respected and most successful, rest on indirect arguments from the observational evidence. New discoveries can always call those arguments into question, showing scientists that the observed data should be understood in a different way, that they have misread their evidence.

But scientists often forget the fallibility of their enterprise. This is not just absentmindedness or wishful thinking. During the heyday of a scientific theory, so much evidence may support the theory, so many observational clues may seem to attest to its truth, that the idea that it could be overthrown appears ludicrous. In addition, the theory may provide ways of identifying quickly what is inaccessible to our unaided senses. Electron microscopes and cloud chambers are obvious examples of those extensions of our perceptual system that theories can inspire. Trained biochemists will talk quite naturally of seeing large molecules, and it is easy to overlook the fact that they are presupposing a massive body of theory in describing what they "see." If that theory were to be amended, even in subtle ways, then the descriptions of the "observed characteristics" of large molecules might have to be given up. Nor should we pride ourselves that the enormous successes of contemporary science secure us against future amendments. No theory in the history of science enjoyed a more spectacular career than Newton's mechanics. Yet Newton's ideas had to give way to Einstein's.

When practicing scientists are reminded of these straightforward points, they frequently adopt what the philosopher George Berkeley called a "forlorn skepticism." From the idea of science as certain and infallible, they jump to a cynical description of their endeavors. Science is sometimes held to be a game played with arbitrary rules, an irrational acceptance of dogma, an enterprise based ultimately on faith. Once we have appreciated the fallibility of natural science and recognized its sources, we can move beyond the simple opposition of proof and faith. Between these extremes lies the vast field of cases in which we believe something on the basis of good—even excellent—but inconclusive evidence.

If we want to emphasize the fact that what scientists believe today may have to be revised in the light of observations made tomorrow, then we can describe all our science as "theory." But the description should not confuse us. To concede that evolutionary biology is a theory is not to suppose that there are alternatives to it that are equally worthy of a place in our curriculum. All theories are revisable, but not all theories are equal. Even though our present evidence does not *prove* that evolutionary biology—or quantum physics, or plate tectonics, or any other theory—is true, evolutionary biologists will maintain that the present evidence is overwhelmingly in favor of their theory and overwhelmingly against its supposed rivals. Their enthusiastic assertions that evolution is a proven fact can be charitably understood as claims that the (admittedly inconclusive) evidence we have for evolutionary theory is as good as we ever obtain for any theory in any field of science.

Hence the Creationist try for a quick Fools' Mate can easily be avoided. Creationists attempt to draw a line between evolutionary biology and the rest of science by remarking that large-scale evolution cannot be observed. This tactic fails. Large-scale evolution is no more inaccessible to observation than nuclear reactions or the molecular composition of water. For the Creationists to succeed in divorcing evolutionary biology from the rest of science, they need to argue that evolutionary theory is less well supported by the evidence than are theories in, for example, physics and chemistry. It will come as no surprise to learn that they try to do this. To assess the merits of their arguments we need a deeper understanding of the logic of inconclusive justification. We shall begin with a simple and popular idea: Scientific theories earn our acceptance by making successful predictions.

Predictive success

Imagine that somebody puts forward a new theory about the origins of hay fever. The theory makes a number of startling predictions concerning connections that we would not have thought worth investigating. For example, it tells us that people who develop hay fever invariably secrete a particular substance in certain fatty tissues and that anyone who eats rhubarb as a child never develops hay fever. The theory predicts things that initially appear fantastic. Suppose that we check up on these predictions and find that they are borne out by clinical tests. Would we not begin to believe—and believe reasonably—that the theory was *at least* on the right track?

This example illustrates a pattern of reasoning that is familiar in the history of science. Theories win support by producing claims about what can be observed, claims that would not have seemed plausible prior to the advancement of the theory, but that are in fact found to be true when we make the appropriate observations. A classic (real) example is Pascal's confirmation of Torricelli's hypothesis that we live at the bottom of an ocean of air that presses down upon us. Pascal reasoned that if Torricelli's hypothesis were true, then air pressure should decrease at higher altitudes (because at higher altitudes we are closer to the "surface" of the atmosphere, so that the length of the column of air that presses down is shorter). Accordingly, he sent his brother-in-law to the top of a mountain to make some barometric measurements. Pascal's clever working out of the observational predictions of Torricelli's theory led to a dramatic predictive success for the theory.

The idea of predictive success has encouraged a popular picture of science. (We shall see later that this picture, while popular, is not terribly accurate.) Philosophers sometimes regard a theory as a collection of claims or statements. Some of these statements offer generalizations about the features of particular, recondite things (genes, atoms, gravitational force, quasars, and the like). These statements are used to infer statements whose truth or falsity can be decided by observation. (This appears to be just what Pascal did.) Statements belonging to this second group are called the *observational consequences* of the theory. Theories are supported when we find that their observational consequences (those that we have checked) are true. The credentials of a theory are damaged if we discover that some of its observational consequences are false.

We can make the idea more precise by being clearer about the inferences involved. Those who talk of inferring observational predictions from our theories think that we can *deduce* from the statements of the theory, and from those statements alone, some predictions whose accuracy we can check by direct observation. Deductive inference is well understood. The fundamental idea of deductive inference is this: We say that a statement S is a valid deductive consequence of a group of statements if and only if it is *impossible* that all the statements in the group should be true and that S should be false; alternatively, S is a valid deductive consequence (or, more simply, a valid consequence) of a group of statements if and only if it would be self-contradictory to assert all the statements in the group and to deny S.

It will be helpful to make the idea of valid consequence more familiar with some examples. Consider the statements "All lovers of baseball dislike George Steinbrenner" and "George Steinbrenner loves baseball." The statement "George Steinbrenner dislikes himself" is a deductively valid consequence of these two statements. For it is impossible that the first two should be true and the third false. However, in claiming that this is a case of deductively valid consequence, we do not commit ourselves to maintaining that *any* of the statements is true. (Perhaps there are some ardent baseball fans who admire Steinbrenner. Perhaps Steinbrenner himself has no time for the game.) What deductive validity means is that the truth of the first two statements would guarantee the truth of the third; that is, *if* the first two *were* true, then the third would have to be true.

Another example will help rule out other misunderstandings. Here are two statements: "Shortly after noon on January 1, 1982, in the

Oval Office, a jelly bean was released from rest more than two feet above any surface"; "Shortly after noon on January 1, 1982, in the Oval Office, a jelly bean fell." Is the second statement a deductively valid consequence of the first? You might think that it is, on the grounds that it would have been impossible for the unfortunate object to have been released and not to have fallen. In one sense this is correct, but that is not the sense of impossibility that deductive logicians have in mind. Strictly speaking, it is not *impossible* for the jellybean to have been released without falling; we can imagine, for example, that the law of gravity might suddenly cease to operate. We do not *contradict* ourselves when we assert that the jellybean was released but deny that it fell; we simply refuse to accept the law of gravity (or some other relevant physical fact).

Thus, *S* is a deductively valid consequence of a group of statements if and only if there is *absolutely no possibility* that all the statements in the group should be true and *S* should be false. This conception allows us to state the popular view of theory and prediction more precisely. Theories are collections of statements. The observational consequences of a theory are statements that have to be true if the statements belonging to the theory are all true. These observational consequences also have to be statements whose truth or falsity can be ascertained by direct observation.

My initial discussion of predictive success presented the rough idea that, when we find the observational consequences of a theory to be true, our findings bring credit to the theory. Conversely, discovery that some observational consequences of a theory are false was viewed as damaging. We can now make the second point much more precise. Any theory that has a false observational consequence must contain some false statement (or statements). For if all the statements in the theory were true, then, according to the standard definitions of *deductive validity* and *observational consequence*, any observational consequence would also have to be true. Hence, if a theory is found to have a false observational consequence, we must conclude that one or more statements of the theory is false.

This means that theories can be conclusively falsified, through the discovery that they have false observational consequences. Some philosophers, most notably Sir Karl Popper (Popper 1959; 1963), have taken this point to have enormous significance for our understanding of science. According to Popper, the essence of a scientific theory is that it should be *falsifiable*. That is, if the theory is false, then it must be possible to show that it is false. Now, if a theory has utterly no observational consequences, it would be extraordinarily difficult to unmask that theory as false. So, to be a genuine scientific theory, a group of statements must have observational consequences. It is important to realize that Popper is not suggesting that every good theory must be false. The difference between being falsifiable and being false is like the difference between being vulnerable and actually being hurt. A good scientific theory should not be false. Rather, it must have observational consequences that could reveal the theory as mistaken if the experiments give the wrong results.

While these ideas about theory testing may seem strange in their formal attire, they emerge quite frequently in discussions of science. They also find their way into the creation-evolution debate.

Predictive failure

From the beginning, evolutionary theory has been charged with just about every possible type of predictive failure. Critics of the theory have argued that (a) the theory makes no predictions (it is unfalsifiable and so fails Popper's criterion for science), (b) the theory makes false predictions (it is falsified), (c) the theory does not make the kinds of predictions it ought to make (the observations and experiments that evolutionary theorists undertake have no bearing on the theory). Many critics, including several Creationists (Morris 1974a; Gish 1979; Wysong 1976), manage to advance all these objections in the same work. This is somewhat surprising, since points (a) and (b) are, of course, mutually contradictory.

The first objection is vitally important to the Creationist cause. Their opponents frequently insist that Creationism fails the crucial test for a scientific theory. The hypothesis that all kinds of organisms were separately fashioned by some "originator" is unfalsifiable (Gould 1981b). Creationists retort that they can play the same game equally well. *Any* hypothesis about the origins of life, including that advanced by evolutionary theory, is not subject to falsification. Hence we cannot justify a decision to teach evolutionary theory and not to teach Creationism by appealing to the Popperian criterion for genuine science.

The allegation that evolutionary theory fails to make any predictions is a completely predictable episode in any Creationist discussion of evolution. Often the point is made by appeal to the authority of Popper. Here are two sample passages:

The outstanding philosopher of science, Karl Popper, though himself an evolutionist, pointed out cogently that evolution, no less than creation, is untestable and thus unprovable. (Morris 1974b, 80)

Thus, for a theory to qualify as a scientific theory, it must be supported by events, processes or properties which can be observed, and the theory must be useful in predicting the outcome of future natural phenomena or laboratory experiments. An additional limitation usually imposed is that the theory must be capable of falsification. That is, it must be possible to conceive some experiment, the failure of which would disprove the theory.

It is on the basis of such criteria that most evolutionists insist that creation be refused consideration as a possible explanation for origins. Creation has not been witnessed by human observers, it cannot be tested experimentally, and as a theory it is nonfalsifiable.

The general theory of evolution also fails to meet all three of these criteria, however. (Gish 1979, 13)

These passages, and many others (for example, Morris 1974a, 150; Morris 1975, 9; Moore 1974, 9; Wilder-Smith 1981, 133), draw on the picture of science sketched above. It is not clear that the Creationists really understand the philosophical views that they attempt to apply. Gish presents the most articulate discussion of the falsifiability criterion. Yet he muddles the issue by describing falsifiability as an "additional limitation" beyond predictive power. (The previous section shows that theories that make predictions are automatically falsifiable.) Nevertheless, the Creationist challenge is a serious one, and, if it could not be met, evolutionary theory would be in trouble.

Creationists buttress their charge of unfalsifiability with further objections. They are aware that biologists frequently look as though they are engaged in observations and experiments. Creationists would allow that researchers in biology sometimes make discoveries. What they deny is that the discoveries support evolutionary theory. They claim that laboratory manipulations fail to teach us about evolution in nature: "Even if modern scientists should ever actually achieve the artificial creation of life from non-life, or of higher kinds from lower kinds, in the laboratory, this would not *prove* in any way that such changes did, or even could, take place in the past by random natural processes" (Morris 1974a, 6). The standards of evidence to be applied to evolutionary biology have suddenly been raised. In this area of inquiry, it is not sufficient that a theory yield observational consequences whose truth or falsity can be decided in the laboratory. Creationists demand special kinds of predictions, and will dismiss as irrelevant any laboratory evidence that evolutionary theorists produce. [In this way, they try to defend point (c).]

Oddly enough, however, the most popular supplement to the charge that evolutionary theory is unfalsifiable is a determined effort to falsify it [point (b)]. Creationists cannot resist arguing that the theory is actually falsified. Some of them, Morris and Gish, for example, recognize the tension between the two objections. They try to paper over the problem by claiming that evolutionary theory and the Creationist account are both "models." Each "model" would "naturally" incline us to expect certain observational results. A favorite Creationist ploy is to draw up tables in which these "predictions" are compared. When we look at the tables we find that the evolutionary expectations are confounded. By contrast, the Creationist "model" leads us to anticipate features of the world that are actually there. Faced with such adverse results, the benighted evolutionary biologist is portrayed as struggling to "explain away" the findings by whatever means he can invent.

Morris's own practice of this form of evolution baiting can serve as an example. Morris constructs a table (1974a, 12; see facing page) whose function is to indicate "the predictions that would probably be made in several important categories" (1974a, 12). Morris admits magnanimously that "these primary models may be modified by secondary [additional] assumptions to fit certain conditions. For example, the basic evolution model may be extended to include harmful, as well as beneficial, mutations, but this is not a natural prediction of the basic concept of evolution" (1974a, 13). The idea that the "natural predictions" of the evolution "model" are at odds with the phenomena is used to suggest that evolutionary biologists are forced to desperate measures to protect their "faith." As Morris triumphantly concludes, "The data must be *explained* by the evolutionist, but they are *predicted* by the creationist" (1974a, 13).

The careful reader ought to be puzzled. If Morris really thinks that evolutionary theory has been falsified, why does he not say so? Of course, he would have to admit that the theory is falsifiable. Seemingly, however, a staunch Creationist should be delighted to abandon a relatively abstruse point about unfalsifiability in favor of a clear-cut refutation. The truth of the matter is that the alleged refutations fail. No evolutionary theorist will grant that (for example) the theory predicts that the fossil record should show "innumerable transitions." Instead, paleontologists will point out that we can deduce conclusions about what we should find in the rocks only if we make assumptions about the fossilization process. Morris makes highly dubious assumptions, hails them as "natural," and then announces that the "natural predictions" of the theory have been defeated.

(This example suggests a method for coping with Morris's "table of natural predictions." Each of these predictions can be deduced from evolutionary theory only if the theory is extended by adding extra assumptions. Morris saddles evolutionary theory with faulty additional

Category	Evolution Model	Creation Model
Structure of Natural Law	Constantly changing	Invariable
Galactic Universe	Galaxies changing	Galaxies constant
Structure of Stars	Stars changing into other types	Stars unchanged
Other Heavenly Bodies	Building up	Breaking down
Types of Rock Formations	Different in different "Ages"	Similar in all "Ages"
Appearance of Life	Life evolving from non-life	Life only from life
Array of Organisms	Continuum of Organisms	Distinct Kinds of Organisms
Appearance of Kinds of Life	New Kinds Appearing	No New Kinds Appearing
Mutations in Organisms	Beneficial	Harmful
Natural Selection	Creative Process	Conservative Process
Age of Earth	Extremely Old	Probably Young
Fossil Record	Innumerable Transitions	Systematic Gaps
Appearance of Man	Ape-Human Intermediates	No Ape-Human Intermediates
Nature of Man	Quantitatively Superior to Animals	Qualitatively Distinct from Animals
Origin of Civilization	Slow and Gradual	Contemporaneous with Man

claims. These are the source of the false predictions. Later, I shall show this in detail for some of Morris's "natural predictions" and the similar difficulties raised by other Creationists [Gish 1979, 53–54; Wysong 1976, 421–426].)

To make a serious assessment of these broad Creationist charges, we must begin by asking some basic methodological questions. We cannot decide whether evolutionary biologists are guilty of trying to save their theory by using ad hoc assumptions (new and implausible claims dreamed up for the sole purpose of protecting some cherished ideas) unless we have some way of deciding when a proposal is ad hoc. Similarly, we cannot make a reasoned response to the charge that laboratory experiments are irrelevant, or to the fundamental objection that evolutionary theory is unfalsifiable, unless we have a firmer grasp of the relation between theory and evidence.

Naive falsificationism

The time has come to tell a dreadful secret. While the picture of scientific testing sketched above continues to be influential among scientists, it has been shown to be seriously incorrect. (To give my profession its due, historians and philosophers of science have been trying to let this particular cat out of the bag for at least thirty years. See, for example, Hempel 1951; Quine 1952.) Important work in the history of science has made it increasingly clear that no major scientific theory has ever exemplified the relation between theory and evidence that the traditional model presents.

What is wrong with the old picture? Answer: Either it debars most of what we take to be science from counting as science or it allows virtually anything to count. On the traditional view of "theory," text-book cases of scientific theories turn out to be unfalsifiable. Suppose we identify Newtonian mechanics with Newton's three laws of motion plus the law of gravitation. What observational consequences can we deduce from these four statements? You might think that we could deduce that if, as the (undoubtedly apocryphal) story alleges, an apple became detached from a branch above where Newton was sitting, the apple would have fallen on his head. But this does not follow at all. To see why not, it is only necessary to recognize that the failure of this alleged prediction would not force us to deny any of the four statements of the theory. All we need do is assume that some other forces were at work that overcame the force of gravity and caused the apple to depart from its usual trajectory. So, given this simple way of applying Popper's criterion, Newtonian mechanics would be unfalsifiable. The same would go for any other scientific theory. Hence none of what we normally take to be science would count as science. (I might note that Popper is aware of this problem and has suggestions of his own as to how it should be overcome. However, what concerns me here are the *applications* of Popper's ideas, that are made by Creationists, as well as by scientists in their professional debates.)

The example of the last paragraph suggests an obvious remedy. Instead of thinking about theories in the simple way just illustrated, we might take them to be far more elaborate. Newton's laws (the three laws of motion and the law of gravitation) are *embedded* in Newtonian mechanics. They form the core of the theory, but do not constitute the whole of it. Newtonian mechanics also contains supplementary assumptions, telling us, for example, that for certain special systems the effects of forces other than gravity are negligible. This more elaborate collection of statements *does* have observational consequences and *is* falsifiable.

But the remedy fails. Imagine that we attempt to expose some self-styled spiritual teacher as an overpaid fraud. We try to point out that the teacher's central message—"Quietness is wholeness in the center of stillness"—is unfalsifiable. The teacher cheerfully admits that, taken by itself, this profound doctrine yields no observational consequences. He then points out that, by themselves, the central statements of scientific theories are also incapable of generating observational consequences. Alas, if all that is demanded is that a doctrine be embedded in a group of statements with observational consequences, our imagined guru will easily slither off the hook. He replies, "You have forgotten that my doctrine has many other claims. For example, I believe that if quietness is wholeness in the center of stillness, then flowers bloom in the spring, bees gather pollen, and blinkered defenders of so-called science raise futile objections to the world's spiritual benefactors. You will see that these three predictions are borne out by experience. Of course, there are countless others. Perhaps when you see how my central message yields so much evident truth, you will recognize the wealth of evidence behind my claim. Quietness is wholeness in the center of stillness."

More formally, the trouble is that *any* statement can be coupled with other statements to produce observational consequences. Given any doctrine D, and any statement O that records the result of an observation, we can enable D to "predict" O by adding the extra assumption, "If D, then O." (In the example, D is "Quietness is wholeness in the center of stillness"; examples of O would be statements describing the blooming of particular flowers in the spring, the pollen gathering of specific bees, and so forth.)

The falsifiability criterion adopted from Popper—which I shall call the *naive falsificationist* criterion—is hopelessly flawed. It runs aground on a fundamental fact about the relation between theory and prediction: On their own, individual scientific laws, or the small groups of laws that are often identified as theories, do not have observational consequences. This crucial point about theories was first understood by the great historian and philosopher of science Pierre Duhem. Duhem saw clearly that individual scientific claims do not, and cannot, confront the evidence one by one. Rather, in his picturesque phrase, "Hypotheses are tested in bundles." Besides ruling out the possibility of testing an individual scientific theory (read, small group of laws), Duhem's insight has another startling consequence. We can only test relatively large bundles of claims. What this means is that when our experiments go awry we are not logically compelled to select any particular claim as the culprit. We can always save a cherished hypothesis from refutation by rejecting (however implausibly) one of the other members of the bundle. Of course, this is exactly what I did in the illustration of Newton and the apple above. Faced with disappointing results, I suggested that we could abandon the (tacit) additional claim that no large forces besides gravity were operating on the apple.

Creationists wheel out the ancient warhorse of naive falsificationism so that they can bolster their charge that evolutionary theory is not a science. The (very) brief course in deductive logic plus the whirlwind tour through naive falsificationism and its pitfalls enable us to see what is at the bottom of this seemingly important criticism. Creationists can appeal to naive falsificationism to show that evolution is not a science. But, given the traditional picture of theory and evidence I have sketched, one can appeal to naive falsificationism to show that *any* science is not a science. So, as with the charge that evolutionary change is unobservable, Creationists have again failed to find some "fault" of evolution not shared with every other science. (And, as we shall see, Creationists like some sciences, especially thermodynamics.) Consistent application of naive falsificationism can show that anybody's favorite science (whether it be quantum physics, molecular biology, or whatever) is not science. Of course, what this shows is that the naive falsificationist criterion is a very poor test of genuine science. To be fair, this point can cut both ways. Scientists who charge that "scientific" Creationism is unfalsifiable are not insulting the theory as much as they think.

Successful science

Despite the inadequacies of naive falsificationism, there is surely something right in the idea that a science can succeed only if it can fail. An invulnerable "science" would not be science at all. To achieve a more adequate understanding of how a science can succeed and how it runs the risk of failure, let us look at one of the most successful sciences and at a famous episode in its development.

Newtonian celestial mechanics is one of the star turns in the history of science. Among its numerous achievements were convincing explanations of the orbits of most of the known planets. Newton and his successors viewed the solar system as a collection of bodies subject only to gravitational interactions; they used the law of gravitation and the laws of motion to compute the orbits. (Bodies whose effects were negligible in any particular case would be disregarded. For example, the gravitational attraction due to Mercury would not be considered in working out the orbit of Saturn.) The results usually tallied beautifully with astronomical observations. But one case proved difficult. The outermost known planet, Uranus, stubbornly followed an orbit that diverged from the best computations. By the early nineteenth century it was clear that something was wrong. Either astronomers erred in treating the solar system as a Newtonian gravitational system or there was some particular difficulty in applying the general method to Uranus.

Perhaps the most naive of falsificationists would have recommended that the central claim of Newtonian mechanics—the claim that the solar system is a Newtonian gravitational system—be abandoned. But there was obviously a more sensible strategy. Astronomers faced one problematical planet, and they asked themselves what made Uranus so difficult. Two of them, John Adams and Urbain Leverrier, came up with an answer. They proposed (independently) that there was a hitherto unobserved planet beyond Uranus. They computed the orbit of the postulated planet and demonstrated that the anomalies of the motion of Uranus could be explained if a planet followed this path. There was a straightforward way to test their proposal. Astronomers began to look for the new planet. Within a few years, the planet— Neptune—was found.

I will extract several morals from this success story. The first concerns an issue we originally encountered in Morris's "table of natural predictions:" What is the proper use of auxiliary hypotheses? Adams and Leverrier saved the central claim of Newtonian celestial mechanics by offering an auxiliary hypothesis. They maintained that there were more things in the heavens than had been dreamed of in previous natural philosophy. The anomalies in the orbit of Uranus could be explained on the assumption of an extra planet. Adams and Leverrier worked out the exact orbit of that planet so that they could provide a detailed account of the perturbations—and so that they could tell their fellow astronomers where to look for Neptune. Thus, their auxiliary hypothesis was *independently testable*. The evidence for Neptune's existence was not just the anomalous motion of Uranus. The hypothesis could be checked independently of any assumptions about Uranus or about the correctness of Newtonian celestial mechanics—by making telescopic observations.

Since hypotheses are always tested in bundles, this method of checking presupposed other assumptions, in particular, the optical principles that justify the use of telescopes. The crucial point is that, while hypotheses are always tested in bundles, they can be tested in *different* bundles. An auxiliary hypothesis ought to be testable independently of the particular problem it is introduced to solve, independently of the theory it is designed to save.

While it is obvious in retrospect—indeed it was obvious at the time—that the problem with Uranus should not be construed as "falsifying" celestial mechanics, it is worth asking explicitly why scientists should have clung to Newton's theory in the face of this difficulty. The answer is not just that nothing succeeds like success, and that Newton's theory had been strikingly successful in calculating the orbits of the other planets. The crucial point concerns the way in which Newton's successes had been achieved. Newton was no opportunist, using one batch of assumptions to cope with Mercury, and then moving on to new devices to handle Venus. Celestial mechanics was a remarkably *unified* theory. It solved problems by invoking the same pattern of reasoning, or *problem-solving strategy*, again and again: From a specification of the positions of the bodies under study, use the law of gravitation to calculate the forces acting; from a statement of the forces acting, use the laws of dynamics to compute the equations of motion; solve the equations of motion to obtain the motions of the bodies. This single pattern of reasoning was applied in case after case to yield conclusions that were independently found to be correct.

At a higher level, celestial mechanics was itself contained in a broader theory. Newtonian physics, as a whole, was remarkably unified. It offered a strategy for solving a diverse collection of problems. Faced with *any* question about motion, the Newtonian suggestion was the same: Find the forces acting, from the forces and the laws of dynamics work out the equations of motion, and solve the equations of motion. The method was employed in a broad range of cases. The revolutions of planets, the motions of projectiles, tidal cycles and pendulum oscillations—all fell to the same problem-solving strategy.

We can draw a second moral. A science should be *unified*. A thriving science is not a gerrymandered patchwork but a coherent whole. Good theories consist of just one problem-solving strategy, or a small family of problem-solving strategies, that can be applied to a wide range of problems. The theory succeeds as it is able to encompass more and more problem areas. Failure looms when the basic problem-solving strategy (or strategies) can resolve almost none of the problems in its intended domain without the "aid" of untestable auxiliary hypotheses.

Despite the vast successes of his theory, Newton hoped for more. He envisaged a time when scientists would recognize other force laws, akin to the law of gravitation, so that other branches of physics could model themselves after celestial mechanics. In addition, he suggested that many physical questions that are not ostensibly about motion— questions about heat and about chemical combination, for example— could be reduced to problems of motion. *Principia*, Newton's master-piece, not only offered a theory; it also advertised a program:

I wish we could derive the rest of the phenomena of Nature by the same kind of reasoning from mechanical principles, for I am induced by many reasons to suspect that they may all depend upon certain forces by which the particles of bodies, by some causes hitherto un-known, are either mutually impelled towards one another, and cohere in regular figures, or are repelled and recede from one another. These forces being unknown, philosophers have hitherto attempted the search of Nature in vain; but I hope the principles here laid down will afford some light either to this or some truer method of philosophy. (Newton 1687/Motte-Cajori 1960, xviii)

Newton's message was clear. His own work only began the task of applying an immensely fruitful, unifying idea.

Newton's successors were moved, quite justifiably, to extend the theory he had offered. They attempted to show how Newton's main problem-solving strategy could be applied to a broader range of physical phenomena. During the eighteenth and nineteenth centuries, the search for understanding of the forces of nature was carried into hydrody-namics, optics, chemistry, and the studies of heat, elasticity, electricity, and magnetism. Not all of these endeavors were equally successful. Nevertheless, Newton's directive fostered the rise of some important new sciences.

The final moral I want to draw from this brief look at Newtonian physics concerns *fecundity*. A great scientific theory, like Newton's, opens up new areas of research. Celestial mechanics led to the discovery of a previously unknown planet. Newtonian physics as a whole led to the development of previously unknown sciences. Because a theory presents a new way of looking at the world, it can lead us to ask new questions, and so to embark on new and fruitful lines of inquiry. Of the many flaws with the earlier picture of theories as sets of statements, none is more important than the misleading presentation of sciences as static and insular. Typically, a flourishing science is incomplete. At any time, it raises more questions than it can currently answer. But incompleteness is no vice. On the contrary, incompleteness is the mother of fecundity. Unresolved problems present challenges that enable a theory to flower in unanticipated ways. They also make the theory hostage to future developments. A good theory should be productive; it should raise new questions and presume that those questions can be answered without giving up its problem-solving strategies.

I have highlighted three characteristics of successful science. *Independent testability* is achieved when it is possible to test auxiliary hypotheses independently of the particular cases for which they are introduced. *Unification* is the result of applying a small family of problem-solving strategies to a broad class of cases. *Fecundity* grows out of incompleteness when a theory opens up new and profitable lines of investigation. Given these marks of successful science, it is easy to see how sciences can fall short, and how some doctrines can do so badly that they fail to count as science at all. A scientific theory begins to wither if some of its auxiliary assumptions can be saved from refutation only by rendering them untestable; or if its problem-solving strategies become a hodgepodge, a collection of unrelated methods, each designed for a separate recalcitrant case; or if the promise of the theory just fizzles, the few questions it raises leading only to dead ends.

When does a doctrine fail to be a science? If a doctrine fails sufficiently abjectly as a science, then it fails to be a science. Where bad science becomes egregious enough, pseudoscience begins. The example of Newtonian physics shows us how to replace the simple (and incorrect) naive falsificationist criterion with a battery of tests. Do the doctrine's problem-solving strategies encounter recurrent difficulties in a significant range of cases? Are the problem-solving strategies an opportunistic collection of unmotivated and unrelated methods? Does the doctrine have too cozy a relationship with auxiliary hypotheses, applying its strategies with claims that can be "tested" only in their applications? Does the doctrine refuse to follow up on unresolved problems, airily dismissing them as "exceptional cases"? Does the doctrine restrict the domain of its methods, forswearing excursions into new areas of investigation where embarrassing questions might arise? If all, or many, of these tests are positive, then the doctrine is not a poor scientific theory. It is not a scientific theory at all.

The account of successful science that I have given not only enables us to replace the naive falsificationist criterion with something better. It also provides a deeper understanding of how theories are justified. Predictive success is one important way in which a theory can win our acceptance. But it is not the only way. In general, theories earn their laurels by solving problems—providing answers that can be independently recognized as correct—and by their fruitfulness. Making a prediction is answering a special kind of question. The astronomers who used celestial mechanics to predict the motion of Mars were answering the question of where Mars would be found. Yet, very frequently, our questions do not concern *what* occurs, but *why* it occurs. We already know that something happens and we want an explanation. Science offers us explanations by setting the phenomena within a unified framework. Using a widely applicable problem-solving strategy, together with independently confirmed auxiliary hypotheses, scientists show that what happened was to be expected. It was known before Newton that the orbits of the planets are approximately elliptical. One of the great achievements of Newton's celestial mechanics was to apply its problem-solving strategy to deduce that the orbit of any planet will be approximately elliptical, thereby explaining the shape of the orbits. In general, science is at least as concerned with reducing the number of unexplained phenomena as it is with generating correct predictions.

The most global Creationist attack on evolutionary theory is the claim that evolution is not a science. If this claim were correct, then the dispute about what to teach in high school science classes would be over. In earlier parts of this chapter, we saw how Creationists were able to launch their broad criticisms. If one accepts the idea that science requires proof, or if one adopts the naive falsificationist criterion, then the theory of evolution—and every other scientific theory—will turn out not to be a part of science. So Creationist standards for science imply that there is no science to be taught.

However, we have seen that Creationist standards rest on a very poor understanding of science. In light of a clearer picture of the scientific enterprise, I have provided a more realistic group of tests for good science, bad science, and pseudoscience. Using this more sophisticated approach, I now want to address seriously the global

Creationist questions about the theory of evolution. Is it a pseudo-science? Is it a poor science? Or is it a great science? These are very important questions, for the appropriateness of granting equal time to Creation "science" depends, in part, on whether it can be regarded as the equal of the theory of evolution.

Darwin's daring

The heart of Darwinian evolutionary theory is a family of problem-solving strategies, related by their common employment of a particular style of historical narrative. A *Darwinian history* is a piece of reasoning of the following general form. The first step consists in a description of an ancestral population of organisms. The reasoning proceeds by tracing the modification of the population through subsequent generations, showing how characteristics were selected, inherited, and became prevalent. (As I noted in chapter 1, natural selection is taken to be the primary—but not the only—force of evolutionary change.)

Reasoning like this can be used to answer a host of biological questions. Suppose that we want to know why a contemporary species manifests a particular trait. We can answer that question by supplying a Darwinian histⱦ ·y that describes the emergence of that trait. Equally, we can use Darwinian histories to answer questions about relationships among groups of organisms. One way to explain why two species share a common feature is to trace their descent from a common ancestor. Questions of biogeography can be addressed in a similar way. We can explain why we find contemporary organisms where we do by following the course of their historical modifications and migrations. Finally, we can tackle problems about extinction by showing how characteristics that had enabled organisms to thrive were no longer advantageous when the environment (or the competition) changed. In all these cases, we find related strategies for solving problems. The history of the development of populations, understood in terms of variation, competition, selection, and inheritance, is used to shed light on broad classes of biological phenomena.

The questions that evolutionary theory has addressed are so numerous that any sample is bound to omit important types. The following short selection undoubtedly reflects the idiosyncrasy of my interests: Why do orchids have such intricate internal structures? Why are male birds of paradise so brightly colored? Why do some reptilian precursors of mammals have enormous "sails" on their backs? Why do bats typically roost upside down? Why are the hemoglobins of humans and apes so similar? Why are there no marsupial analogues of seals and whales? Why is the mammalian fauna of Madagascar so distinctive? Why did the large, carnivorous ground birds of South America become extinct? Why is the sex ratio in most species one to one (although it is markedly different in some species of insects)? Answers to these questions, employing Darwinian histories, can be found in works written by contemporary Darwinian biologists. Those works contain answers to a myriad of other questions of the same general types. Darwinian histories are constructed again and again to illuminate the characteristics of contemporary organisms, to account for the similarities and differences among species, to explain why the forms preserved in the fossil record emerged and became extinct, to cast light on the geographical distribution of animals and plants.

We can see the theory in action by taking a brief look at one of these examples. The island of Madagascar, off the east coast of Africa, supports a peculiar group of mammals. Many of these mammals are endemic. Among them is a group of relatively small insectivorous mammals, the *tenrecs*. All tenrecs share certain features that mark them out as relatively primitive mammals. They have very poor vision, their excretory system is rudimentary, the testes in the male are carried within the body, their capacity for regulating their body temperature is poor compared with that of most mammals. Yet, on their simple and rudimentary body plan, specialized characteristics have often been imposed. Some tenrecs have the hedgehog's method of defense. Others have the forelimbs characteristic of moles. There are climbing tenrecs that resemble the shrews, and there are tenrecs that defend themselves by attempting to stick their quills into a would-be predator. Hedgehogs, moles, tree shrews, and porcupines do not inhabit Madagascar. But they seem to have their imitators. (These are examples of convergent evolution, cases in which unrelated organisms take on some of the same characteristics.) Why are these peculiar animals found on Madagascar, and nowhere else?

A straightforward evolutionary story makes sense of what we observe. In the late Mesozoic or early Cenozoic, small, primitive, insectivorous mammals rafted across the Mozambique channel and colonized Madagascar. Later the channel widened and Madagascar became inaccessible to the more advanced mammals that evolved on the mainland. Hence the early colonists developed without competition from advanced mainland forms and without pressure from many of the normal predators who make life difficult for small mammals. The tenrecs have been relatively protected. In the absence of rigorous competition, they have preserved their simple body plan, and they have exploited unoccupied niches, which are filled elsewhere by more advanced creatures. Tenrecs have gone up the trees and burrowed in the ground because those are good ways to make a living and because they have had nobody but one another to contend with.

The same kind of story can be told again and again to answer all sorts of questions about all sorts of living things. Evolutionary theory is unified because so many diverse questions—questions as various as those I listed—can be addressed by advancing Darwinian histories. Moreover, these narratives constantly make claims that are subject to independent check. Here are four examples from the case of the triumphant tenrecs. (1) The explanation presupposes that Madagascar has drifted away from the east coast of Africa. That is something that can be checked by using geological criteria for the movement of land-masses, criteria that are independent of biology. (2) The account claims that the tenrecs would have been able to raft across the Mozambique channel, but that the present channel constitutes a barrier to more advanced mammals (small rodents). These claims could be tested by looking to see whether the animals in question can disperse across channels of the appropriate sizes. (3) The narrative assumes that the specialized methods of defense offered advantages against the predators that were present in Madagascar. Studies of animal interactions can test whether the particular defenses are effective against local predators. (4) Central to the explanatory account is the thesis that the tenrecs are related. If this is so, then studies of the minute details of tenrec anatomy should reveal many common features, and the structures of proteins ought to be similar. In particular, the tenrecs ought to be much more like one another than they are like hedgehogs, shrews, or moles.

Looking at one example, or even at a small number of examples, does not really convey the strength of evolutionary theory. The same patterns of reasoning can be applied again and again, in book after book, monograph after monograph, article after article. Yet the particular successes in dealing with details of natural history, numerous though they are, do not exhaust the accomplishments of the theory. Darwin's original theory—the problem-solving strategies advanced in the *Origin*, which are, in essence, those just described—gave rise to important new areas of scientific investigation. Evolutionary theory has been remarkably fruitful.

Darwin not only provided a scheme for unifying the diversity of life. He also gave a structure to our ignorance. After Darwin, it was important to resolve general issues about the presuppositions of Darwinian histories. The way in which biology should proceed had been made admirably plain, and it was clear that biologists had to tackle questions for which they had, as yet, no answers. How do new characteristics arise in populations? What are the mechanisms of inheritance? How do characteristics become fixed in populations? What criteria decide when a characteristic confers some advantage on its possessor? What interactions among populations of organisms affect the adaptive value of characteristics? With respect to all of these questions, Darwin was forced to confess ignorance. By raising them, his theory pointed the way to its further articulation.

Since Darwin's day, biologists have contributed parts of evolutionary theory that help to answer these important questions. Geneticists have advanced our understanding of the transmission of characteristics between generations and have enabled us to see how new characteristics can arise. Population geneticists have analyzed the variation present in populations of organisms; they have suggested how that variation is maintained and have specified ways in which characteristics can be fixed or eliminated. Workers in morphology and physiology have helped us to see how variations of particular kinds might yield advantages in particular environments. Ecologists have studied the ways in which interactions among populations can affect survival and fecundity.

The moral is obvious. Darwin gambled. He trusted that the questions he left open would be answered by independent biological sciences and that the deliverances of these sciences would be consistent with the presuppositions of Darwinian histories. Because of the breadth of his vision, Darwin made his theory vulnerable from a number of different directions. To take just one example, it could have turned out the mechanisms of heredity would have made it impossible for advantageous variations to be preserved and to spread. Indeed, earlier in this century, many biologists felt that the emerging views about inheritance did not fit into Darwin's picture, and the fortunes of Darwinian evolutionary theory were on the wane.

When we look at the last 120 years of the history of biology, it is impossible to ignore the fecundity of Darwin's ideas. Not only have inquiries into the presuppositions of Darwinian histories yielded new theoretical disciplines (like population genetics), but the problem-solving strategies have been extended to cover phenomena that initially appeared troublesome. One recent triumph has been the development of explanations for social interactions among animals. Behavior involving one animal's promotion of the good of others seems initially to pose a problem for evolutionary theory. How can we construct Darwinian histories for the emergence of such behavior? W. D. Hamilton's concept of inclusive fitness, and the deployment of game-theoretic ideas by R. L. Trivers and John Maynard Smith, revealed how the difficulty could be resolved by a clever extension of traditional Darwinian concepts.

Yet puzzles remain. One problem is the existence of sex. When an organism forms gametes (sperm cells or egg cells) there is a meiotic division, so that in sexual reproduction only half of an organism's genes are transmitted to each of its progeny. Because of this "cost of meiosis," it is hard to see how genotypes for sexual reproduction might have become prevalent. (Apparently, they will spread only half as fast as their asexual rivals.) So why is there sex? We do not have a compelling answer to the question. Despite some ingenious suggestions by orthodox Darwinians (notably G. C. Williams 1975; John Maynard Smith 1978), there is no convincing Darwinian history for the emergence of sexual reproduction. However, evolutionary theorists believe that the problem will be solved without abandoning the main Darwinian insights—just as early nineteenth-century astronomers believed that the problem of the motion of Uranus could be overcome without major modification of Newton's celestial mechanics.

The comparison is apt. Like Newton's physics in 1800, evolutionary theory today rests on a huge record of successes. In both cases, we find a unified theory whose problem-solving strategies are applied to illuminate a host of diverse phenomena. Both theories offer problem solutions that can be subjected to rigorous independent checks. Both open up new lines of inquiry and have a history of surmounting apparent obstacles. The virtues of successful science are clearly displayed in both.

There is a simple way to put the point. Darwin is the Newton of biology. Evolutionary theory is not simply an area of science that has had some success at solving problems. It has unified biology and it has inspired important biological disciplines. Darwin himself appreciated the unification achieved by his theory and its promise of further development (Darwin 1859/Mayr 1964, 188, 253–254, 484–486). Over a century later, at the beginning of his authoritative account of current views of species and their origins, Ernst Mayr explained how that promise had been fulfilled: "The theory of evolution is quite rightly called the greatest unifying theory in biology. The diversity of organisms, similarities and differences between kinds of organisms, patterns of distribution and behavior, adaptation and interaction, all this was merely a bewildering chaos of facts until given meaning by the evolutionary theory" (Mayr 1970, 1). Dobzhansky put the point even more concisely: "Nothing in biology makes sense except in the light of evolution" (Dobzhansky 1973).

3

Darwin Redux

The tautology objection

Consistency is not the hobgoblin of the Creationist mind. However, Creationist motives are not hard to discern. Creationists would like to show that the theory of evolution is simply false. To this end, they hunt diligently for observational findings that would cast doubt on parts of the theory, and they revel in unresolved disputes among evolutionary biologists. On the other hand, the claim that evolution is untestable is essential to their case for equal time. Since "[n]either evolution nor creation is accessible to the scientific method," both can aptly be described as "religion." So both "models of origins" should be taught in the classroom (Morris 1974b, 172–173). Given the strong pull of their objectives, Creationists throw consistency to the winds and try to press both types of criticism.

Because the point about untestability is so vital to their cause, they try to support it in a number of different ways. In the last chapter, we saw how the criterion of naive falsificationism (adapted from Popper) failed to pin any "fault" on evolution that is not shared by every other science. But the appeal to this criterion does not exhaust Creationist efforts.

One popular way to try to argue that evolutionary theory is not testable, that it differs from real science, can be called the *tautology objection*. The central idea is that evolutionary theory reduces to an empty truism. This is another ancient warhorse, which Creationists ride with zeal. Here are some typical Creationist versions:

[Macbeth] points out that although evolutionists have abandoned classical Darwinism, the modern synthetic theory they have proposed as a substitute is equally inadequate to explain progressive change as the result of natural selection, and, as a matter of fact, they cannot even define natural selection in nontautologous terms. (Gish 1979, 20)

Schutzenberger and others have shown the above Neodarwinian approach to biogenesis and the origin of species to be tautological, i.e. meaningless. The reasoning behind Schutzenberger's claim is quite elementary in reality, for he points out that the Neodarwinian hypothesis simply states nothing more than that the organism which survives has survived. Or put otherwise: the organism leaving the greatest number of offspring behind will survive. This type of depth of wisdom is not very difficult to plumb. (Wilder-Smith 1981, 127; see also Morris 1974a, 6–7)

The tautology objection is not an original contribution of the Creationists. Similar arguments have been voiced by philosophers (Popper 1963), lawyers (Macbeth 1971), and biologists (Peters 1976). The objection has even found its way into *Harper's Magazine* (Bethell 1976). Once again, there is good reason to believe that the Creationists do not really understand the objection they are borrowing. It is attractive to them because it ends in the right place. Some of their formulations are downright mystifying however. Tautologies are standardly taken to be the simplest logical truths, for example, statements of the form "All *A*'s are *A*'s," such as "All cats are cats." Logicians ought to be upset by the idea that tautologies are "meaningless" (Wilder-Smith). While a statement like "All cats are cats" is not exactly informative, it is perfectly clear what it means. Still, we can all take comfort in the fact that an "organism leaving the greatest number of offspring *behind will* survive" (Wilder-Smith).

Despite the infelicities of Creationist formulations, it is worth looking more closely at the tautology ojection. The objection has certainly attracted a surprising number of adherents. What accounts for its fascination? There are two sources of confusion. The more important of these is the tendency to boil evolutionary theory down to a single statement: the principle of natural selection. The other is misunderstanding of the concept of Darwinian fitness. Working together, these muddles produce the idea that evolutionary theory reduces to the claim that the fittest survive—and that what this means is that those who survive survive.

I shall begin with the concept of fitness. The first point to note is that fitness is not concerned with survival as an end in itself, but rather with survival as a means to reproduction. From the Darwinian standpoint, what counts for fitness is success in leaving descendants. But if we simply measure the fitness of an organism by the number of offspring it leaves, then peculiar consequences threaten. Organisms that we would naturally regard as equally fit can differ in their fortunes. Let us consider two antelopes, each with the same complement of genes, happily grazing on a plain. A cheetah approaches from one side, seizing the nearer of the two. The other antelope escapes and goes on to produce numerous offspring. Obviously it would be absurd to declare that the two organisms have different degrees of fitness. The inequality in their actual reproductive successes is simply a matter of luck. So biologists say that the fitness of an organism varies as its *expected* reproductive success. What is important is *the number of offspring that an organism with those genes could have been expected to produce.*

What is this number? The last paragraph provides a constraint on any definition of fitness; we want to count two organisms with the same genes as equally fit. Contemporary evolutionary theory provides an approach to fitness designed to satisfy this constraint. The concept of fitness as expected reproductive success is introduced in terms of *genes*, not in terms of *organisms*. Scientists estimate the relative fitnesses of different alleles at the same locus by taking the fitness of an allele (or, more exactly, of an allelic pair) to be measured by its representation in future generations (see, for example, Wilson and Bossert 1971, 47–54). Here it is possible to use the *actual* reproductive success as an estimate because the same allelic pair will be found in many different organisms, so that the effects of good and bad luck are expected to cancel. Insofar as one can speak of the fitness of the organism—its expected reproductive success—it will be a function of the fitnesses of the allelic pairs which make up its genotype. (I should note that there are some obvious shortcomings of this approach to fitness. Genes do not act singly, but in combination. Hence it is something of an abstraction to identify the fitness of an allelic pair, in isolation, and then to regard the fitness of the organism as determined by the fitnesses of its allelic pairs. However, for some purposes, the abstraction works well enough.)

Now that the concept of fitness is somewhat clearer, it is time to ask what the principle of natural selection actually says about the fittest. The obvious way to answer this question would be to consult a presentation of evolutionary theory and to find the statement marked "Principle of Natural Selection" (just as we might look to a textbook on mechanics for the "Principle of Inertia"). Surprisingly, this approach yields no answer. The indexes of two excellent and thorough textbooks in evolutionary theory (Dobzhansky, Ayala, Stebbins, and Valentine 1977; Futuyma 1979) have lengthy entries for natural selection, but *no* entry for the *principle* of natural selection. That is, the phrase "natural selection" occurs many times in presenting evolutionary biology, but there is no one general statement given pride of place as the principle of natural selection.

However, there is one part of evolutionary theory—mathematical population genetics—that offers very definite conclusions about fitness and representation in future generations. It is not easy to formulate the idea that "the fittest survive" if we try to provide a principle about *organisms*. On the other hand, if we want a principle about *genes*, then we can obtain detailed mathematical results about the relations between fitness and the representation of genotypes. The task of mathematical population genetics is to work out the mathematical intricacies of the ways in which the distributions of genetic combinations will vary in a sequence of populations, according to the relative fitnesses of the relevant allelic pairs. (Consider a simple example. Suppose that, at a locus, there are two alleles, A and a. The allelic pairs AA, Aa, aa have relative fitnesses 0.9, 1.0, and 0. It is a theorem of population genetics that the lethal allele a will persist in the population at a frequency of 0.09. The case of sickle-cell anemia, described in chapter 1, is similar to this.)

Mathematical population genetics articulates precisely the idea that genes that are more fit become prevalent in a population. It offers detailed descriptions of how the relative fitnesses of alleles will result in the spread of allelic combinations, how under some conditions equilibria will be reached, and so forth. The claim that the fittest survive becomes an array of definite results about the distribution of genes in successive populations.

So the tautology objection is wrong in the first place because the principle of natural selection is not a tautology. Insofar as the principle has an heir in contemporary evolutionary theory, it is a collection of theorems in mathematical population genetics. But there is a more significant error. We should not identify the Darwinian theory of evolution with the principle of natural selection to begin with. Fundamental to the tautology objection is the false philosophical view scotched in the last chapter, the view that any theory can be identified with a small collection of statements. This view is applied to evolutionary theory to generate the caricature that the theory is just the principle of natural selection.

Let us approach the issue in terms of the positive account offered in the last chapter. Evolutionary theory is a collection of problem-solving strategies that use Darwinian histories. In constructing Darwinian histories, the concept of fitness is constantly employed. Darwinian histories frequently make claims about factors that promote or detract from fitness. One theorist, who is investigating cryptic coloring (protective camouflage) in moths, points out that an ability to blend with its environment makes an organism less vulnerable to predators. Another, who is interested in the physiognomy of bats, identifies the upturning of the nose as assisting the organism in its use of sound waves to locate neighboring objects. In these, and countless other cases, biologists use the concept of fitness and *make independently testable claims about what gives the organisms in question whatever fitness they have.*

The example of cryptic coloring in moths provides an especially good demonstration of how claims about the grounds of fitness can be independently evaluated. Naturalists can watch predators at work. They can record the relative numbers of captures involving well-camouflaged moths and those that stand out from their surroundings. Moreover, they can eliminate other possible reasons for differential reproductive success. For example, it is possible to show that protective coloration makes no difference in fecundity, survival of larvae, or ability to mate. Thus they confirm the claim that the success results from the protection afforded by cryptic coloration.

The main point is that the *concepts* of fitness and natural selection, not the *principle* of natural selection, play a central role in Darwinian evolutionary biology. Darwin made some relatively elementary observations. Organisms vary. The variation leads to differential abilities for survival and reproduction. Some of the characteristics that underlie differences in reproductive success can be inherited. What Darwin saw was that these elementary observations can give rise to problem-solving strategies of enormous potential. They suggest a general pattern of evolutionary explanation. When evolutionary theorists give explanations, they make statements about the fitness of particular organisms with particular characteristics, statements that are vulnerable to independent checks. They do not simply intone, "The fittest survive."

In the last chapter, I pointed out some defects of the traditional view that theories are small collections of statements. Whatever attractiveness that view may have when we think about Newtonian mechanics, it becomes ludicrous when we attempt to apply it to sciences like genetics, plate tectonics, and evolutionary theory. These sciences employ distinctive *concepts*—the concepts of gene, plate, and fitness— and they deploy these concepts in distinctive ways to solve problems. But there are no general laws or principles involving the concepts that can be picked out as the whole (or even the essence) of any of these theories. Geologists use the concept of plate (and of interactions among plates) to explain earthquakes (among other things). They do not simply respond to a particular earthquake by stating some "principle of plates."

Rather, they advance testable claims about the motions of the plates in the case at hand. Analogously, no reputable biologist accounts for an evolutionary development by asserting the principle of natural selection—as if it were enough to say "It's the survival of the fittest again." Instead, each time the concept of fitness is employed in evolutionary explanations, biologists are compelled to advance some independent, empirically vulnerable, claim about the advantage conferred by a particular characteristic under particular circumstances.

The principle of natural selection is not a tautology. Even if it were, it would not follow that evolutionary theory is tautologous and untestable. To generate that resounding conclusion, we would have to reduce evolutionary theory to the principle of natural selection. Since the reduction is hopelessly wrong, the objection collapses. Whatever Creationists may think, evolutionary biologists are not in the habit of declaiming *ad nauseam* that those who survive survive.

Anything goes?

Here is still another way of trying to bolster the charge that evolutionary theory is untestable. The claim is that evolutionary theory is full of fantasizing. It consists of Just-So Stories without Kipling's wit. Once again, let us hear from the Creationists:

Furthermore, the architects of the modern synthetic theory of evolution have so skillfully constructed their theory that it is not capable of falsification. The theory is so plastic that it is capable of explaining anything. (Gish 1979, 17)

[Evolutionary theory] bridges enormous gaps in the fossil record with creatures of the imagination. (Watson 1979, 97)

Rather than evolution being science, the exposé of, the objections to, the criticisms are science [sic]. Scientific facts, laws and principles must be dismissed or seriously altered and tampered with to maintain one's belief in evolution. (Wysong 1976, 460)*

As with the tautology objection, the accusation is not new. Darwin's earliest critics objected that evolutionary theory is so accommodating that it can adapt to any observational finding. Fleeming Jenkin's presentation is forceful, specific and eloquent:

*Strictly speaking, these sentences are uttered by Wysong's Creationist spokesman. However, by the late stage in the book at which this passage appears, the pretence of an impartial author moderating a debate has long been dropped. So I have no qualms about attributing them directly to Wysong.

The peculiarities of geographical distribution seem very difficult of explanation on any theory. Darwin calls in alternately, winds, tides, birds, beasts, all animated nature, as the diffusers of species, and then a good many of the same agencies as impenetrable barriers. There are some impenetrable barriers between the Galapagos Islands, but not between New Zealand and South America. Continents are created to join Australia and the Cape of Good Hope, while a sea as broad as the Bristol Channel is elsewhere a valid line of demarcation. With these facilities of hypothesis there seems no particular reason why many theories should not be true. However an animal may have been produced, it must have been produced somewhere, and it must either have spread very widely, or not have spread, and Darwin can give good reason for both results. (Jenkin 1867; in Hull 1974, 342)

The charges are serious. They portray the pronouncements of evolutionary theory as on a par with the "predictions" found in fortune cookies. What is wrong with such "predictions" is that they can adapt themselves to all circumstances. Plainly, a theory that is compatible with all results explains none. But is it true that in evolutionary theorizing anything goes?

One initial qualification is in order. I have no doubt that we can discover cases in which evolutionary theorists have clutched at unconfirmed assumptions to save cherished hypotheses about the historical development of particular kinds of organisms. Biology is done by humans, and it would be surprising if the desire to protect one's past work did not sometimes override the force of the evidence. However, behavior of this sort does not have to occur. When evolutionary theorists dodge criticisms by inventing ad hoc hypotheses, they are not pursuing a practice endemic to their discipline, but flouting the limits that it imposes. Under these circumstances, they are properly criticized by their scientific colleagues.

Fleeming Jenkin's general worry was that the checks on Darwinian histories are too weak, so that there are no limits on the biologist's imagination. Yet, as we saw in the last chapter, Darwinian histories can be subjected to independent checks. If I hypothesize that two contemporary forms are related to a recent common ancestor, then studies of anatomical and biochemical similarities can test the extent of the relationship. If I suggest that a successful variation could have been fixed in a population in a specific number of generations, then my suggestion is vulnerable to the results of population genetics. If I propose that a modification of structure could have generated a reproductive advantage, then my proposal can be compared with what we know about the relation between similar modifications and reproductive success. Geological evidence can be amassed to test whatever claims I make about continental connections or climatic conditions. My liberty to invent scenarios is constantly checked. (Once again, let me note that if evolutionary theorists have sometimes been allowed to indulge their imaginations about particular evolutionary developments, that is not because their enterprise lacks rules but because they—and their colleagues—have been lax in enforcing the rules.)

Hence there is no basis for the blanket charge that evolutionary theory is a game without rules. As I have emphasized, independently testable hypotheses and theories provide numerous checks on Darwinian histories. Moreover, the proposals of evolutionary theorists provide cross checks on one another. If one scientist attempts to explain the emergence of a group of West Indian lizards and another tries to account for the distribution of birds in the same region, then the hypotheses about climate, faunal diversity, land connections, and so forth must be compatible with one another. As the enterprise of articulating the history of life proceeds, the resolution of outstanding problems will be ever more sharply constrained by the facts already established.

Hence, if there is any substance to the worry that evolutionary theory is too accommodating, it will have to be because there are identifiable loopholes that scientists can exploit or special strategems that allow the imagination free rein. Fleeming Jenkin made it quite clear what he found objectionable. Evolutionary theorists, he claimed, can erect barriers where they wish and make whatever suppositions they please about past connections between continents and islands. But this is incorrect. The thesis that a particular obstacle, such as a body of water, constitutes a barrier for a species of organism can be tested in perfectly straightforward ways. Equally, geological theory enables us to test suggestions about previous land connections. However, to be fair to Jenkin, the resources for checking those suggestions are more adequate now than they were in Darwin's day.

Contemporary Creationists are fond of the broad denunciation that evolutionary theory plays fast and loose with the observations. It is important to ask for the details. Where exactly does the chameleon character of the theory reveal itself? The answer can be gleaned from Creationist writings: Evolutionary theorists stack the deck by defining terms so as to favor their views. For example, the much vaunted geological time scale is set up to fit the claims of the theory. In addition to these all-purpose ways in which evolutionary theory protects itself, there are also particular examples of explaining away troublesome data by ad hoc means. These examples are exposed in Morris's table of "natural predictions." In the next two sections I shall examine these objections in some detail.

Doubts about dating

Henry Morris is the main exponent of Creationist doubts about the ways in which evolutionary theorists date the occurrence of ancestral organisms: "Here is obviously a powerful system of circular reasoning. Fossils are used as the only key for placing rocks in chronological order. The criterion for assigning fossils to specific places in that chronology is the assumed evolutionary progression of life; the assumed evolutionary progression is based on the fossil record so constructed. The main evidence for evolution is the assumption of evolution" (Morris 1974a, 136).

The charge levelled here, and in other places (Morris 1974a, 94–96, 144–145; Gish 1979, 57–59; Wilder-Smith 1981, 103), is that geology and evolutionary biology have far too cozy a relationship. Instead of providing independent checks on Darwinian histories, geology obligingly adjusts its account of the ages of rocks to meet the needs of the biologists who are weaving their narratives around the fossil findings. Hence Morris maintains that evolutionary theory has built-in protection; claims about the sequence in which past organisms occurred are not vulnerable to question by geologists. For geology assumes the sequence of organisms predicted by evolution to order the strata in rock formations. In essence, Morris charges evolutionary theorists with viciously circular reasoning. Central claims about the history of life are defended by appealing to the dates that geologists assign to rock strata. But the geologists' claims rest on evolutionary assumptions about the dates of the organisms fossilized in the rocks. Evolution is defended by assuming evolution.

The charge is based on a misrepresentation of the way in which the geological time scale was actually constructed. The geological time scale was set up before Darwin. It was constructed by scientists who believed in events of special creation (although they would not meet the requirements of contemporary Creationist orthodoxy, since they believed in successive waves of creation). Early nineteenth-century geologists used the *sequence* of fossil forms found throughout the world to correlate strata. What they assumed was that organisms found in lower strata had flourished before those whose remains are buried in higher strata. This assumption bore no commitment to the idea that the later organisms *evolved* from the earlier ones.

Moreover, Morris ought to know that his account misrepresents geological practice in a different way. For he and his fellow Creationists argue vehemently against the use of a wide variety of independent methods for dating rocks that appeal to phenomena of radioactive decay. (I shall look briefly at some of these arguments in chapter 5.) Evidently, they believe that these methods are faulty. However, what is presently at issue is whether there is *any independent* method for assigning dates to rock strata. Morris asserts that fossils are "the only key" for ordering rocks in time. That is simply false, and he knows it. Rocks can be dated by using a family of radiometric methods, as well as the techniques of classical stratigraphy. To claim that these methods do not succeed is to offer a different criticism. One cannot simply ignore independent methods and then claim that they do not exist.

In fact, Morris cites a passage that tells directly against the point he wishes to make. He quotes a sensible sentence by three geologists (Stieff, Stern, and Eichler, cited in Morris 1974a, 144–145) to the effect that the best assessment of ages should take into account the results of all the available techniques—including paleontological evidence. Their judgment should remind us of two important points. First, there are many independent ways of assigning ages to rocks (or of telling which rocks preceded which others). Second, when geologists are interested in fixing the age of a rock stratum as precisely as possible, they will standardly take into account the deliverances of all applicable methods.

Here is an analogy. Imagine that you are watching an event that takes a long time to occur and that you want to measure as accurately as you can the time at which it finishes. You have a number of different clocks available to you, and you establish in advance that they are synchronized. Would it not be prudent to use all of the clocks in making your measurement, and to arrive at your result by taking each of them into account? After all, we know that any individual clock can go wrong. By employing several of them, we are able to identify discordant results, and thus protect ourselves against error.

This analogy applies directly to the enterprise of dating rock strata. There are geological methods (the methods of classical stragigraphy) for dividing rock formations into strata and for correlating strata in different locations. These methods provide one way of assigning (relative) ages. Radiometric methods offer another way to attribute ages to rocks. The fossil record also enables geologists to correlate strata, matching rocks that contain similar fossils, and thus helps in the enterprise of constructing a chronological order. However, as I have pointed out, use of fossils does not even presuppose evolutionary theory; all that is assumed is that there is a correlation between the sequence of fossil forms and the sequence of strata. We can compare the time scales (or chronological orderings) produced by these various techniques and show them to be, by and large, in agreement. Then, when we want to date a particular rock stratum it is appropriate to do what the practicing geologists recommend: Consult all of the available techniques.

This practice is typical of natural science. Frequently, scientists have many ways to measure a particular quantity or to ascertain whether a particular condition is present. The methods are correlated, and, in cases where special accuracy is required, decisions are reached by using a consensus of the available techniques. For example, chemists test for acidity in many different ways, and physicists employ a variety of techniques to measure temperature.

The case of temperature measurement provides further insight into the practice of dating rocks. There are all kinds of thermometers: simple mercury and gas thermometers, thermocouples, instruments that read off temperature in response to the radiation spectra emitted, and a host of even more complicated devices. Over large ranges of temperature, these measuring instruments can be compared and their readings correlated. Yet it is quite obvious that we are not free to use any instrument in any context: If we want to measure the internal temperature of a star, we cannot use a mercury thermometer; nor will spectral analysis give us a good reading of the body temperature of a sick child. Different thermometers are appropriate to different cases. But, because the thermometers *agree* in those instances where they can be compared, we take them all to record temperature measurements on a single scale. Equally, in dating rock strata, we cannot always use every method of assigning an age. Many rocks lack the radioactive decay products that allow us to apply radiometric methods. For some rocks the classical stratigraphic techniques are inapplicable. Yet other rocks do not contain fossils, and hence cannot be dated by the fossils they bear. What geologists do is exactly what physicists do. Having established the agreement of various methods of measurement, they assign dates to rock strata by whatever methods can be applied to the case at hand.

Seen in this light, the use of "index fossils," which Creationists roundly condemn, is a perfectly harmless procedure. There are some rock strata that bear distinctive fossils but do not lend themselves to precise stratigraphic analysis or to radiometric dating. Geologists date those strata by using the correlation of the fossil sequence with the stratigraphic order and the radiometric scale. The fossils in question are interpreted as the remains of organisms that flourished at a particular time, and that time is taken as the date of the strata. "Circularity!," cry the Creationists. But there is no circle here. The use of the fossil record to date rock strata is justified by the *independent correlation* of the time scales constructed using stratigraphic, radiometric, and biological methods. Once that correlation has been established, we can use *any* of the correlated methods. Where only one method is available, that method will naturally be employed.

Creationists attempt to show that methods of dating rocks insulate evolutionary theory against geological counterevidence. But their argument rests on a distortion of geological practice. Once the character of that practice is revealed, it turns out to be exactly the same as the methods used by sciences that do not find themselves berated by critics with special interests.

Unnatural predictions

It is now time to return to Morris's "table of basic predictions." As we discovered in chapter 2, Morris, Gish, and other Creationists claim that, although evolutionary theory makes no genuine predictions (since it is unfalsifiable), it does make "basic predictions." That is, the "evolution model" suggests that certain things will be found in nature. Since we do not find these things, evolutionary theorists resort to fudging. The "natural predictions" fail, so that "secondary assumptions" have to be introduced. Or so the Creationists charge.

In this section, I shall consider three alleged "natural predictions" of evolutionary theory. Two of these figure in Morris's table: evolutionary theory "predicts"that the "structure of natural laws" will be "constantly changing," and there will be a "continuum of organisms" with no "distinct kinds." The third is a traditional theme, which Creationists (including Morris) sometimes emphasize. Evolution "predicts" that the temporal sequence of organisms will show an "upward development." Inexplicably, this particular chestnut does not find a place on Morris's table. I shall examine at this stage only these three alleged "basic predictions." However, an additional item from the table will occupy us in chapter 4, since the question of the fossil record requires a more extensive discussion.

Let us start with one of Morris's tastier red herrings. According to Morris, "It seems obvious that the evolution model would predict that matter, energy and the laws are still evolving since they must have evolved in the past and there is no external agent to bring such evolution to a halt" (1974a, 18). But, as he tells us triumphantly, the laws of nature do not evolve: "That is, the law of gravity, the laws of thermodynamics, the laws of motion and all other truly basic laws have apparently always functioned in just the way they do now, contrary to a prediction of the basic evolution model" (Morris 1974a, 18). What do defenders of evolutionary theory do when they find themselves in this difficult predicament? Morris's answer is that they resort to special pleading: "These stable aspects of nature can of course be accommodated within the evolution model, but only at the cost of introducing a secondary assumption therein—namely, that the laws completed their own evolution at some time in the past and have been stable since. The point is that this situation *requires explanation* in the framework of the evolution model. The creation model, on the other hand, does not have to explain it,—it predicts it!" (Morris 1974a, 19). Of course, we are supposed to conclude that evolutionary theory protects itself against damaging counterevidence by introducing ad hoc assumptions.

What is most puzzling about this argument is that anybody would ever believe that evolutionary theory favors the evolution of laws of nature. Evolutionary theory is not committed to some sort of pan-evolutionism—as if to believe that *some* things (such as species) evolve were to believe that *everything* (including the laws of nature) evolves. A moment's reflection, however, will lead us to see that evolutionary theory itself assumes the *invariance* of at least some laws of nature, the laws of chemical combination, for example. Further, there is a simple logical point. If there is to be a theory of the evolution of organisms, then the task of that theory will be to specify the laws that govern the successive emergence of forms of life (principles of biochemistry, for example). Those laws must be taken to remain constant throughout the evolutionary process. In a nutshell, the concept of a law governed evolutionary development requires that the governing laws are not themselves part of the evolutionary development.

Apart from this logical point, there are a few simple facts. The cosmological theory within which evolutionary theory is embedded— the cosmology that encompasses the development of the universe from those first explosive minutes—includes the basic laws of physics that Morris mentions. Those laws are regarded as invariant through time. The task of the grand cosmological version of evolutionary theory is to apply them to presumptive initial conditions to account for the formation of elements, nebulae, galaxies, the solar system, the earth, and, ultimately, biologically significant molecules on the earth. At that stage, it is relevant to appeal to the (biological) evolutionary theory with which I am chiefly concerned. Nowhere in the theory is the commitment to temporal invariance of laws of nature abandoned. What *is* true is that different laws of nature figure in our explanations at different points in the grand account. Quite evidently, scientists will only start appealing to principles of biochemistry when they begin to explain the interactions of biologically significant molecules. But this is not to suppose that the laws of biochemistry evolve, that new laws, which were not previously true, suddenly become true. The laws of biochemistry are timelessly true. Just as the law of falling bodies remains true even when no bodies are falling, so too the laws that govern objects formed relatively late in the development of the cosmos were true from the very beginning.

Cosmology tells us how different kinds of things—from inorganic molecules up to higher organisms—were successively formed. In telling the grand story, scientists appeal to different sciences in succession. This is not because the laws of nature evolve, however, but because the nature they are describing evolves. Once these simple points have been appreciated, we see that evolutionary theorists were never forced to make any arbitrary decision about "when the laws of nature stopped evolving." Evolving laws are creatures of Morris's imagination.

The second alleged "prediction" of the "evolution model" I wish to discuss is the thesis that living forms should make up a continuum. Once again, Morris is the principal exponent of the criticism, although other Creationists echo the point (Hiebert 1979, 52):

If an evolutionary continuum existed, as the evolution model should predict, there would be no gaps, and thus it would be impossible to demark specific categories of life. Classification requires not only similarities, but differences and gaps as well, and these are much more amenable to the creation model. (Morris 1974a, 72)

Gaps and differences are not predicted at all by the evolution model, except on the basis of subsidiary hypotheses that must be introduced for this specific purpose. Thus the very existence of a science of taxonomy is a prediction of creationism and a problem to evolutionism. (Morris 1974b, 85)

In order to isolate the supposed "subsidiary hypotheses," let us ask how evolutionary theory explains the discontinuity of living forms. Why does evolutionary theory represent the history of life as a tree with separated branches? There are a number of answers that a neo-Darwinian might offer. In some cases intermediate forms may simply not be produced. An ancestral population may split into two parts· that can then diverge from one another by natural selection. After millions of years of evolution the two descendant populations are morphologically very different, and there are no intermediates. In other cases, intermediates may arise but lose out in competition with the extreme forms. Neither of these explanations is exactly an unrelated straw frantically clutched by Darwinians to save their favorite theory.

I shall support this evaluation by considering the second scenario, in which distinct species emerge from a competitive process involving the defeat of intermediates. There is a genuine point behind the Creationist criticism. For what is really demanded is that Darwinians fill in the details about how speciation occurs. Notoriously, *The Origin of Species* is rather inexplicit about the origin of species. Now, as I noted in the last chapter, Darwin's program could succeed only if a large number of details were filled in (in appropriate ways) by a large number of different sciences. One place in which Darwin's ambitious theory was vulnerable was on the question of speciation. And, once again, Darwin's promise was redeemed. Contemporary evolutionary theorists, most notably Mayr (1963; 1970) have specified the type of Darwinian history that should be constructed for cases in which a new species splits off from an ancestral population.

Having presented the broad outline of the theory of speciation in chapter 1, I shall fill in some needed details via a hypothetical example. An evolutionary ornithologist discovers two related species of birds. Among other things, they differ conspicuously in the average lengths of their beaks. Let us imagine that in one species the average length of beak is 3 centimeters; in the other, 6 centimeters. The ornithologist regards the species as having evolved from a common ancestral population. What type of Darwinian history should we construct to describe the evolutionary process? In particular, are special, unsupported assumptions required to explain that discontinuity? These are the serious questions that lie behind the Creationist challenge.

The main tradition of contemporary evolutionary theory (lucidly presented in Mayr 1970) offers the following answers. The first requirement for speciation to occur is that a small group of organisms should become geographically isolated from the rest of the population. This small group may be subjected to environmental pressures not felt by the main body of the population. Because of these pressures, variations from forms that are prevalent in the ancestral population can become adaptively advantageous. If it is successful, the initially isolated minority may give rise to a flourishing descendant population, which is reproductively isolated and morphologically different from the ancestral stock.

The general scenario is easily illustrated by our hypothetical example. Our ornithologist may propose that the short-beaked form represents the ancestral type. (I should emphasize, however, that the claim that both species had a recent common ancestor does not imply that *either* contemporary population represents the ancestral species. It is quite possible that the original species should have been modified in two different ways.) The long-beaked form arose from a small isolated subgroup of the ancestral population that found itself in an environment without a handy supply of worms. In this environment, birds with longer beaks were more efficient predators. Thus there was continuous selection in favor of those birds in the population that had the longest beaks. Our ornithologist suggests that the history of the new species consists of a sequence of populations through which average beak length consistently increased. At some stage in that history, birds with beaks measuring 5 centimeters would have been at a competitive advantage. But, at later stages, those of their descendants who retained the parental form of a 5-centimeter beak were less effective at securing food than their beakier siblings who had varied further from the ancestral type. Each intermediate value represents a characteristic that was advantageous in competition with the other birds present at the time. Yet, once the species achieved that characteristic, it moved on to further advantages. At some point, the variation may be optimal for the environment and the competition. For example, the 6-centi-meter form may not simply represent the latest stage in a process of selection for longer beaks. It may be the best length, in that it enables those birds that realize it to obtain an ample supply of worms without incurring an unnecessary expense of energy. (Smaller beaks may be less efficient. Larger beaks may require extra expenditure of resources to no gain.) If these conditions are met, the intermediate forms (with beaks between 3 and 6 centimeters) will tend to disappear because they are inferior in competition for food. More extravagantly beaked forms have no feeding advantage, and suffer competitive inferiority by squandering developmental resources. Thus we find a stable pop-ulation with an average beak length of 6 centimeters.

Two assumptions underlie the neo-Darwinian account of (geographic) speciation, which I have briefly rehearsed and illustrated. They are neither implausible nor unconfirmed. The first assumption is that a portion of an ancestral population can become isolated in an envi-ronment that makes distinctive demands. Second, neo-Darwinians ap-peal to the commonplace point that not every adaptive variation in a particular environment is equally advantageous. Far from being desperate hypotheses, cooked up to save evolutionary theory from refutation, these claims are about as well confirmed as any statement a biologist is ever likely to make. Investigations of the distributions of organisms teach us that, in many groups of organisms, the geographical isolation of small populations in distinctive environments is (from the perspective of geological time) an everyday affair. (However, this should not lead us to conclude that speciation is occurring with astronomical frequency. Most of the "evolutionary experiments" that go on in isolated populations end in failure.) Moreover, we recognize immediately that a variation can be more advantageous than the original type, even though a further variation would yield an even more effective competitor. To take a different example of the modification of characteristics, crude attempts at camouflage can have adaptive significance for variant moths, but their descendants may be able to achieve a far more perfect cryptic coloration. This second assumption is actually a corollary of a fundamental idea of evolutionary theory: Reproductive success depends on the competition.

What is exciting about the neo-Darwinian account of geographic speciation is not the relatively straightforward idea that evolution by natural selection can produce new kinds, but the ambitious suggestion that speciation always does proceed in the way I have sketched. Using the approach to theories I presented in the last chapter, we can understand orthodox neo-Darwinism as specifying a problem-solving strategy. When we confront a case of speciation, we are to advance a particular kind of Darwinian history, one which involves a claim that the daughter species emerged from an isolated subgroup of the ancestral population. Currently, this orthodoxy is being questioned (by, for example, White 1977). Nevertheless, contemporary disputes offer no aid and comfort to the Creationist cause. Nobody can dispute that *some* cases of speciation are to be understood in the traditional way. The novel suggestion is that evolutionary theory has *more* resources for explaining species diversity than orthodox neo-Darwinism allows.

I turn now to my last version of the charge that evolutionary biologists fudge when their "natural predictions" are confounded. According to Creationists, evolutionary theory is committed to progressivism. Wysong presents the argument as follows: "Prediction: Beneficial organs should be maintained in a population. If man is the acme of evolution, he should have carried with him the most beneficial characteristics" (Wysong 1976, 345). Wysong continues with a series of advertisements for the prowess of nonhumans. Humans turn out to be rather deficient at pulling, seeing, smelling, running, tracking, and working hard. Wysong's Creationist spokesman concludes with some rhetorical questions: "Why did the human not retain such obviously beneficial characteristics in his evolution? Why do lowly creatures outdo us in so many areas of ability? Why did we 'leave behind' beneficial charac-

teristics in our evolutionary ascent?" (Wysong 1976, 346). The impli-
cation is that evolutionary theorists can only answer these questions
by "explaining away the facts," appealing to special assumptions that
have no independent confirmation. Wilder-Smith tries the same gambit:

Under favorable environmental conditions, rats reproduce very rapidly.
With the help of their prolific reproduction they place more highly
developed species at a disadvantage, which is certainly not very con-
ducive to the realization of Neodarwinian evolutionary concepts. Primitive moss easily displaces more highly developed grass in a lawn.
These facts are so obvious that further specific examples are super-
fluous. It is evident that prolific reproduction and the production of
a greater number of offspring are certainly not equivalent to an upward
development and evolution of species, so that we can safely pigeonhole
this particular Neodarwinian concept. (Wilder-Smith 1981, 128)

Both versions link evolution to the concept of progress. Wysong
makes the connection in a simple and crude way: Evolutionary theory
implies that later organisms should be able to perform better in all
respects than their ancestors did. Wilder-Smith is less definite. He is
content merely to mention "upward development" and "evolution"
in the same breath. In both cases, the argument gains power from a
popular idea. If the evolution of species takes place by the fixation of
characteristics that are adaptively advantageous, then surely (living)
things must be getting better all the time.

There have been some serious attempts to work out this idea in
detail. The goal would be to describe, in precise terms, the kind of
progress that we can expect the evolutionary process to produce.
George Williams provides a thorough survey of these attempts, and
argues cogently that none of them succeeds in identifying an adequate
notion of evolutionary progress (Williams 1966, 34–55). Williams states
his thesis very clearly at the beginning of his discussion: "I would
maintain, however, that there is nothing in the basic structure of the
theory of natural selection [that is, the theory of evolution by natural
selection] that would suggest the idea of any kind of cumulative progress" (Williams 1966, 34). Certainly, the theory of evolution by natural
selection does not imply the crude sort of progress discussed by Wysong.
Nor does it predict any kind of "upward development" that would
be contradicted by the behavior of rats and moss to which Wilder-
Smith alludes.

From the evolutionary perspective, successful organisms are op-
portunists. Darwinian histories sometimes describe how newly arising
variations gave their possessors particular advantages in *particular* en-

vironments. But this does not imply that the variations were perfectly designed to achieve the advantages, nor that they would have been advantageous in any environment, nor that they were obtainable without loss of other advantageous characteristics. *Overall*, the new variations were advantageous *in that environment*. It is quite conceivable, however, that under different circumstances they would have proved costly. When environments favor organisms with a particular structure or a particular behavior, it is perfectly possible, even likely, that the successful organisms that realize the structure (exhibit the behavior) may have to forego characteristics that would have been highly advantageous, given other circumstances and other demands.

Let us take a close look at a couple of Wysong's examples of the superior prowess of nonhumans. Some animals run much faster than humans do. The animals cited by Wysong (such as cheetahs) are not direct ancestors of *Homo sapiens*, so that their superior performance does not support the idea that humans have lost a beneficial characteristic possessed by their predecessors. In fact, we understand quite well why humans are not as swift as some carnivores (cheetahs, tigers, wolves) or some of their prey (antelope). Many carnivores and their prey have developed the primitive mammalian hindlimb to make it specialized for running. By contrast, the primates retained the ability of the primitive mammalian hindlimb to serve a large number of purposes. In particular, evolution of the primates has modified the hindlimb under the pressure of a forest environment. Our ancestors were selected not only for their ability to run, but also for their ability to climb. We inherit characteristics that limit our speed on the ground, characteristics that were once valuable in an environment that is no longer our natural habitat. (In fact, humans are probably faster runners than our direct ancestors.)

One ability that is clearly less developed in humans than in our mammalian ancestors is our sense of smell. In this case, we can support Wysong's contention that *Homo sapiens* has lost a characteristic that was beneficial to our progenitors. Yet this does not shake the foundations of evolutionary theory. It is not difficult to construct an evolutionary scenario that will show how we might have lost olfactory acuity. Here is one possibility. In a forest environment, detection of objects by smell seems to be far less reliable than it is on the open plains. Hence, among the arboreal animals from whom we descend, there may have been selection for an alternative method of recognizing the threats and promises of the environment. Vision has become our dominant sense. As our ancestors perfected their ability to gain information by sight, perhaps they no longer found it necessary to expend resources on developing an elaborate olfactory system. So our sense of smell is less acute than was that of the early mammals from whom we began.

All this is characteristic of evolution. Because of the pressures of the current environment, certain features of the organism are favored, even though developing those features may inhibit the retention of other properties that would be advantageous under different environments. It may even be that the properties that are "traded away" would be valuable in environments that descendant populations subsequently encounter. No doubt it would have been useful to those of our ancestors who descended from the trees to run away quickly when predators threatened. Perhaps there were occasions on which some of them would have been saved by a better sense of smell. But they had to do the best they could with equipment that had formerly been developed for arboreal living. The evolutionary process is sometimes shortsighted.

Wysong's criticism rests on a distortion of evolutionary theory. Darwinians never claim that a characteristic is *absolutely* beneficial. Rather, a property is beneficial (or advantageous) in a particular environment. Typically, gaining an advantage requires the organism to give up something that might be useful under other circumstances. So, in looking at the world, there is no surprise in finding that there are many types of organisms, each of which is extraordinarily good at one kind of thing. (Of course, the range of abilities and the types of specialization vary among organisms.) Evolutionary theory is not committed to the absurd idea that successful organisms are supremely good at everything. Nature provides many ways of making a living.

As I have already remarked, Wilder-Smith is less specific in connecting evolution and "upward development." Yet a simple extension of the ideas of the last paragraph should convince us that evolutionary theory is not committed to denying that rats can multiply prolifically and moss can invade a lawn. The fact that one species of organism is more complex than another, or that it emerges at a later stage of evolutionary history, does not mean that it will inevitably succeed in competition in *any* environment. A highly developed species may have acquired characteristics that were advantageous in a limited range of environments. Placed in a different environment, it may not withstand competition from a species that is much less developed. It may even be true that the *absence* of that competitor was crucial to the development of the species. Hence the fact that gardeners have to struggle to prevent moss from encroaching on their lawns is of no significance at all for evolutionary theory. (Exactly parallel considerations apply to the case of the rats that threaten their evolutionary "betters.") We should not even be surprised if the plants whose aesthetic qualities we value are competitively inferior to large numbers of wild species. After all, our favorite plants, shrubs, and grasses are developed in very special environments. Wilder-Smith seems to believe that there is a theorem of evolutionary theory stating that weeding is unnecessary. Since I am an avid gardener, for once I wish that he were right.

We have inspected three specific examples of the charge that evolutionary theory is compelled to appeal to implausible special assumptions when its "natural predictions" go awry. In each case, we reach the same conclusion. Creationists foist off on evolutionary theory some assumptions that are clearly false. Then they proclaim that the theory makes some false predictions. When we uncover the false assumptions, we see that the allegedly "natural predictions" of evolutionary theory are highly *un*natural. Or, to put the point another way, when the problem-solving strategies of the theory are presented correctly, we find that those problem-solving strategies, supplemented with independently confirmed hypotheses, make predictions diametrically opposed to those that Creationists pin on the "evolution model." These objections, and others like them, rest on distortions of the theory under attack. It is not necessary to go very deeply into evolutionary theory to expose the misrepresentations; I have not had to reformulate the theory in some ingenious and precise way. However, the connections alleged by Creationists are likely to sound natural enough to people who have only a superficial acquaintance with evolutionary theory. Morris's table of "basic predictions," as well as the kindred devices used by his fellow Creationists, are well adapted to deceive the nonspecialist's eye.

Disallowing evidence

Only one of the methodological issues canvassed in chapter 2 remains to be tackled. Creationists make the surprising claim that, in the case of evolution, laboratory evidence does not count. Here again is Morris's presentation of the claim: "Even if modern scientists should ever actually achieve the artificial creation of life from non-life, or of higher kinds from lower kinds, in the laboratory, this would not *prove* in any way that such changes did, or even could, take place in the past by random natural processes" (Morris 1974a, 6). Creationists are interested in undercutting the significance of laboratory experiments so that they can buttress a main plank in their platform: "The subject of origins is ultimately beyond the scope of empirical science" (Morris 1974b, 80).

At least two misleading words figure in Morris's formulation. We have already seen that *proof* is not the issue in science; what we are interested in is good evidence. In the next chapter, we shall look at the ways in which Creationists exploit the idea of *randomness*. To focus the issue, I shall abstract from these irrelevant complications. I shall examine one of the experiments Morris explicitly mentions and ask whether this experiment provides *good evidence* for claims about what *natural processes* occurred, or could have occurred, in the past.

Imagine that you are a researcher on the origins of life. Like H. Urey, S. Miller, and their successors, you want to show how it is possible for biologically significant molecules (such as amino acids) to have been formed from much simpler molecules (hydrogen, methane, ammonia, and so forth) that might have occurred in the earth's primitive atmosphere. You cleverly concoct a mixture of simple molecules, simulating a primitive atmosphere. You pass your simulated lightning discharges through it and find, to your delight, that you have all the amino acids, short protein chains, even segments of DNA, that anyone could wish for. Given this (hypothetical) story of (unprecedented) success, what are you entitled to conclude?

We should grant Morris one point. Unless you have significant extra information, one successful experiment does not justify the inference that the particular conditions of the simulated atmosphere correspond to the state present when life began. For to be warranted in that conclusion you would require reasons to believe that *other* concoctions of a "primitive atmosphere" and other combinations of discharges would not deliver the goods. (Actually, you could make the stronger claim if you could eliminate the apparently possible alternatives.) However, if you are a researcher on the origins of life, your chief interest is not in the strong claim that the mixture in your container exactly duplicates what the earth was like at that time. You have a general theoretical view about how life on earth came to be — the contemporary evolutionary view — and you want to resolve a particular question that arises. How could it all have begun? How was the transition from simple molecules to biologically significant molecules possible? Is our general theoretical view on the right track? So your task is to identify a *possible* way in which biologically significant molecules could first have been formed.

Hence, the relevant part of Morris's criticism is his denial that you are entitled to infer that life *could* have originated naturally in the way it did in your experiment. We must be completely clear about what the conclusion is supposed to be. You want to claim that *if* a particular set of conditions had actually been present on the earth's surface several billion years ago, *then* biological molecules would naturally have developed. What you do in your laboratory is to create exactly these conditions—and lo!, the appropriate molecules are formed. You bring about the conditions and let nature take its course. You now conclude that if, by some natural means, these conditions had actually been present on the earth's surface, then nature would have taken its course to generate the biological molecules.

Your work might arouse specific criticisms. Some might object that circumstances on the earth's surface would be relevantly different from those in the laboratory; perhaps the scale of the experiment affects the outcome. Others might worry that the conditions you chose are not realistic; maybe it is hard to see how they could have existed on the earth's surface at that time, so that your result is irrelevant to the question of how biological molecules might have been generated. Such concrete criticisms would have to be addressed by appealing to physical theory and, perhaps, performing further experiments. However, if these specific kinds of criticisms do not arise, or if you can overcome them, then you have good reason to claim that you have solved your problem. You have given a possible account of the origin of life.

Morris believes that there is a *general* objection to this type of inference. What could it be? What is the difference between the kind of inference I have described and the inferences performed by scientists in all fields every day? Finding that certain events regularly follow certain conditions that they bring about in the laboratory, scientists conclude that the regularity holds in nature. (Unless, of course, there are specific reasons for thinking that there is an important difference between the laboratory situation and its natural analogue.)

These questions are answered in forthright discussions by Morris and some of his fellow Creationists. Here is Wilder-Smith on the topic of the significance of research into the origins of biologically significant molecules: "Today nobody any longer attempts to create life from matter and energy only. Nobody places the simple material components of life in a mixer, or stirring machine, thus adding nonteleonomic energy until life is formed. This type of nonsense has not been carried out since the days of Pasteur. Today energy *and* know-how (information, concept, logos) are always added. Since this step has been taken (i.e. know-how has been added), scientists have become successful in their attempts to create artificial life" (Wilder-Smith 1981, 7; see also Wilder-Smith 1970, 138–139; Wysong 1976, 230–231; Morris 1974a, 49). This will come as news to many. Wilder-Smith conjures up the delightful picture of Urey, Miller, and their successors secretly mixing the missing ingredient KNOW-HOW into their "primeval atmosphere." Alas, this picture is thoroughly confused. Experiments that attempt to simulate the origin of life are no different in kind from a host of other chemical experiments, including those we all fondly remember from our high school days. The (more or less) knowing experimenter places certain chemicals in a container. The reaction proceeds without further interference. (Otherwise it is fraud.) The mere fact that the conditions were actually brought about by design does not prevent us from concluding that the same reaction would have proceeded if those very conditions had occurred naturally. There is no mysterious effect, resulting from the KNOW-HOW of the experimenter.

One can only wonder how Wilder-Smith would like laboratory experiments to be run. Should we avoid contamination by KNOW-HOW by keeping chemicals in unlabeled jars? Perhaps there should be a requirement that workers don blindfolds before entering the laboratory? Or we could permit only those who are ignorant of chemistry to design experiments? The discussion of this issue by Wilder-Smith (and parallel passages by Morris and Wysong) reveals why members of the scientific community can be excused for not taking "scientific" Creationists seriously. There are many Creationist objections that, when submitted to serious scrutiny, are simply ridiculous.

The biologist's lot

According to Gilbert and Sullivan, "The policeman's lot is not an 'appy one." If so, policemen and evolutionary theorists have something in common. Since 1859 evolutionary biology has had to face an unusually large number of methodological criticisms. Creationists delight in dredging up hoary objections. We have seen that, when examined, these objections are, at best, misguided; at worst, nonsensical. But suspicion may linger. Where there is so much gossip, surely something scandalous is afoot. Why has evolutionary theory caused so much distrust? In the final chapter, we shall look at one source of worry about the theory—the allegations that it has disastrous implications for religion and morality. But here I want to address a different aspect of the problem. Apart from any social ramifications, why has the theory attracted so many doubts about its scientific status? I shall conclude this chapter by trying to understand why misguided methodological criticisms are part of the evolutionary biologist's lot.

We may compare the situation of physicists specializing in classical mechanics. Any question about the scientific status of classical mechanics can be quickly resolved by using the problem-solving strategies of the theory to make some dramatic prediction. So, for example, physicists can compute the flight path of a rocket that is about to be launched. By doing so, they demonstrate at one stroke the great power of the problem-solving strategies of classical mechanics. Can the evolutionary biologists do anything similar? And if not, why not?

The most spectacular prediction that an evolutionary biologist could make would be to tell us about the future large-scale evolution of some current group of organisms. All methodological objections to evolutionary biology would immediately be silenced if an evolutionary theorist could apply the problem-solving strategies of the theory to tell us what the characteristics of the descendants of the primates will be 10 million years hence. Unfortunately, that cannot be done. However, it ought to be obvious *why* it cannot be done: We simply have no way of knowing all the things that we would have to know to apply our problem-solving strategies to this particular problem. There is nothing vague or indefinite about the problem-solving strategies themselves. *If* we knew enough about the ways in which environments will change, *if* we knew enough about the genetics of organisms—not only the primates themselves but the animals and plants with which they interact—*then* perhaps we could predict the future evolutionary path of the primates. But the supposition that we do—or should—know enough is clearly absurd. Hence the chance to achieve a coup for evolutionary theory eludes us.

Why is the physicist so much luckier? The answer is that at least some of the systems that classical mechanics describes are very simple. To compute the trajectory of our hypothetical rocket, one only has to consider (at least initially) two bodies: the earth and the rocket itself. The system consisting of those two bodies can be studied in happy neglect of other factors. Everyday events will not disturb it. Of course, the system is not completely insulated from external happenings. Some cataclysm in the solar system might affect it quite radically, and then all our bets would be off. But, unless something highly improbable happens, we can focus our attention on the behavior of two bodies alone.

Quite obviously, evolutionary biologists are not so fortunate. When they try to solve the problem of how a species will evolve, almost anything is potentially relevant. A small environmental change could transmit a shock wave through the entire ecosystem, imposing un-anticipated demands on the species under study. Important questions ramify endlessly. Thus the difference between the biologist's and the physicist's lot can be summarized concisely. At least some of the systems that physicists study can be treated as simple systems that are, for all practical purposes, indifferent to whatever occurs outside them. Life — or, more exactly, the development of living things — is not like that. (Even the physicist is not always so lucky. There are numerous intractable cases of many-body problems in classical mechanics.)

Yet evolutionary theory *can* make predictions, albeit of a less spectacular sort. One type of case is the prediction of the evolutionary development of populations under laboratory conditions. Another is the prediction of some features of the past.

Now the idea of predicting the past may sound peculiar. After all, surely a *pre*diction must concern the future? Names do not matter here. What is important is that evolutionary theory does something analogous to prediction — it is sometimes called "postdiction" or "retrodiction" — by telling us about the past. The essence of prediction lies in the fact that a prediction gives us reason to believe a claim that we previously had no reason to believe. (We may even have had reason not to believe it.) Whether the claim concerns the past or the future is irrelevant.

Paleontologists make predictions all the time. They predict where fossils of various types are likely to have been buried. (Field paleontologists do not just set out hopefully, hammer in hand, to tap whatever rock catches their eyes!) They predict the paths that animals traversed in reaching their present geographical distribution. A striking example of the latter kind is the prediction that marsupial mammals once lived in what is now Antarctica. (Antarctica was once adjacent both to South America and to Australia. According to the standard account, it served the marsupials as a bridge in their migration from South America to Australia.) This prediction was recently confirmed by fossil findings. Paleontological prediction is possible because scientists sometimes have enough evidence to apply the problem-solving strategies of evolutionary theory to puzzles about the past. The evidence becomes available as the history of life is disclosed and scientists learn what continental configurations were once like and what variations once occurred.

There is a quite different area in which evolutionary theory can make predictions. By placing organisms whose genetics is well understood in a carefully controlled environment, biologists can emulate the fortunate physicists. Some microorganisms that have a short generation time and whose genetics is known lend themselves to evolutionary predictions. (The rapidity of generations is important, in that the environment needs to be controlled over a significant number of generations.) The following kind of experiment is certainly possible — indeed experiments like it have been performed by Joshua Lederberg and his associates. Scientists can select a strain of some well-understood microorganism that is unable to manufacture a particular protein. They can then choose a means of inducing mutations that is calculated to affect particular regions in the genetic material (including regions where mutation might generate a variant able to produce the protein). The initial setup for the experiment would be to place the strain in an environment where production of the protein is highly advantageous and to treat it with the chosen agent for inducing mutations. Evolutionary theory could then be applied to predict the development of the population.

In both these cases, it is possible to use the problem-solving strategies of the theory to make predictions because the information that is needed for the application is available. However, in many cases, including those that are of most interest, the information is not ready to hand. Hence, to emphasize the predictive power of evolutionary theory is somehow to miss the point, like stressing Ted Williams's prowess as a fisherman or Winston Churchill's talents as an artist. The primary function of the theory is to advance our understanding of past and present organisms, revealing to us how the features of the organic world can be comprehended by recognizing its history. Evolutionary theory merits a place among the sciences for the reasons I have advanced in the past two chapters. It offers a unified set of problem-solving strategies that can be applied, by means of independently testable assumptions, to answer a myriad of questions about the characteristics of organisms, their interrelationships, and their distributions. It is also remarkable for its fecundity. Indeed, evolutionary theory has spawned so many healthy new sciences that its actual reproductive success is truly spectacular. Small wonder, then, that it has survived the long dull litany of methodological discontent.

4

Mountains, Molehills, and Misunderstandings

Grades of grousing

The road to Creationism is paved with bad philosophy. However, the engines that transport us down that road are fueled by bits of science, variously chopped, twisted, crushed, mangled, and blended. Creationists delight in appealing to other parts of science to condemn evolutionary theory. Having examined the methodological criticisms of evolutionary theory, we shall now take up substantive criticisms, that purport to show that well-established facts tell against the theory. No readable chapter could explore all the misconceptions of all the "scientific criticisms" that Creationists perpetrate. We shall investigate only the chief deficiencies of the most important arguments. However, that will enable us to evaluate the Creationists' track record—and so to judge whether some unexamined argument is likely to contain a devastating criticism.

A taxonomy of criticisms will prove useful. Creationist grousing comes in grades. Some complaints are very broad and general, denying the possibility that evolutionary theory can resolve any problem. Others are much more specific, attempting to demonstrate that a certain phenomenon cannot be understood in the ways that the theory allows. Obviously, we have a continuum of criticisms, ranging from the most global to the most local.

Evolutionary theory constructs Darwinian histories to resolve biological questions. The most global criticism of the theory is that none of these problem-solving strategies can ever be implemented because the idea of a Darwinian history—any Darwinian history—has presuppositions that can be shown to be false. For example, Darwinian histories assume that it is possible for higher degrees of organization to arise by random processes; yet the second law of thermodynamics (a law dear to the hearts of the "scientific" Creationists) denies the possibility. Similarly, any Darwinian history must make some claim about the occurrence of favorable mutations. However, if the experiments on mutant fruit flies performed in genetics laboratories over the last sixty years have shown anything, it is that mutations are harmful. Hence there are scientific demonstrations that Darwinian histories cannot be constructed.

Some Creationists advance these global objections to claim a decisive refutation of evolutionary theory. Thus Hiebert maintains (correctly) that evolutionary theory must postulate a mechanism for the development of new species from old. He continues, "In their desperation to find a mechanism, scientists have grasped at mutations, the one agency in nature that causes inheritable deterioration in living organisms! This total inability of scientists to explain how evolution occurred constitutes a major crisis in the continued survival of the theory of evolution" (Hiebert 1979, 35–36—first sentence boldface in text). Wilder-Smith also claims outright victory: "The laws of physics—the laws of thermodynamics—also contradict evolutionary theory. . . . The scientific, experimental evidence which we possess thus speaks decisively against Darwin's theory of evolution" (Wilder-Smith 1981, 6).

However, other Creationists seem willing to allow that evolutionary biologists may find ways of resolving the contradictions between their theory and well-confirmed laws. In his initial discussion of the laws of thermodynamics in *The Troubled Waters of Evolution*, Morris portrays evolutionary theorists as having "faith" that their "model" will overcome the critical problem posed by the "entropy principle" (Morris 1974b, 101). Later, he warms to his task in the manner more usual among Creationists: "The only escape from this dilemma is finally to realize that the whole evolutionary concept can be nothing but a great delusion" (Morris 1974b, 132).

Another objection, which is less global, but still very broad, is the charge that there is too little time for large-scale evolution to have occurred. If correct, this charge would permit the construction of *some* Darwinian histories. However, it would greatly restrict the class of cases to which evolutionary problem-solving strategies can be applied, thus damaging the unifying power of the theory. One particular benefit from the Creationist point of view would be to block the idea of tracing the emergence of contemporary forms from some primeval soup. This would give Creationism a critical foot in the door. For in this case, Creationists suggest, scientists would have to concede that, if left alone, natural processes would not have sufficed to bring about the origin of life. One would have to appeal to a Creator to explain particular steps in the evolution of living forms. Once this concession is made, and the appeal to the Divine Intervener, whose actions help out natural processes, is allowed to stand, there would be no reason to stop short of full-fledged Creationism. (Although I do not endorse this argument, I think that there is no doubt that Creationist reasoning flows more smoothly if it is admitted that some events are beyond the scope of scientific law.)

Moving along the continuum, we next encounter the objection based on the fossil record. The criticism begins by noting that, in a large range of cases, the Darwinian histories that evolutionary biologists construct postulate a large number of ancestral forms that are not found in the rocks. However, Creationists are uncertain about exactly how they want to press the point. As we have seen, some hope to convict evolutionary theory of insulating itself against all possible refutation by "special pleading" (Morris 1974a, 90). Others, Gish, for example, are more aggressive. They want to show that the fossil record actually refutes evolutionary theory: What the fossils say is "No!" The central message of Gish's book (Gish 1979) is that the fossil record is at odds with so many proposed Darwinian histories that we can consider evolutionary theory to have been falsified. As he puts it at one enthusiastic moment, "The rocks cry out 'Creation!' " (Gish 1979, 76).

More specific Creationist worries concern particular cases (or classes of cases) for which Darwinian histories are supposed to be difficult to find. A favorite version focuses on complex organs and patterns of behavior. How can we give an evolutionary explanation of structures like the eye, the pollen "baskets" of flowers, the stings of wasps, and so forth? Other versions concentrate on very particular phenomena, for which Creationists claim that no evolutionary account can be given. So we are treated to a magical mystery tour. Among the highlights are the Paluxy river bed (where, so the story goes, human footprints traverse dinosaur traces), Rhodesia (where there are cave drawings said to depict dinosaurs), Glacier National Park (where the evolutionists' ordering of rock strata is "inexplicably" reversed), and the coast of Africa (where, to dumbfound evolutionary theorists, an alleged "missing link" between fishes and amphibians turned up alive and well after all these years).

Creationists use scientific theories and scientific observations to present a wide variety of objections to evolutionary theory. Some are Brobdignagian flourishes, attempting to crush the theory in a single blow. Others are Lilliputian pinpricks, detailed points that, in sufficient numbers, could cripple even a giant. Between these extremes lie objections of intermediate scale. My aim will be to descend from the large to the minute, from the general to the specific. On my way I shall try to restore to their original form those scientific ideas that Creationists bludgeon into serving their ends. Since several of the global criticisms employ a common tactic, I shall begin by disarming this weapon of obfuscation.

The randomness ploy

Creationists are fond of what I shall call "the randomness ploy." They delight in noting that evolutionary theory commonly talks of "random mutations" (and sometimes of variations that *chance* to be useful to the organisms exhibiting them or of *accidental* isolation of populations). Then the refrain begins: How could *chance* produce the order of the world of living things?

The following passages are a small sample of those in which Creationists harp on the idea that evolutionary theory portrays the development of life as a chance process:

It seems beyond all question that such complex systems as the DNA molecule could never arise by chance, no matter how big the universe or how long the time. The creation model faces this fact realistically and postulates a great Creator, by whom came life. (Morris 1974a, 62)

A well known principle in physics is the Heisenberg principle of uncertainty. The principle basically argues that the movement of electrons and atoms is a random process. If the atom is governed by random processes, how could there be biochemical bias? If the law of the atom is randomness how can we cite the atom as the source of order? (Wysong 1976, 126)

But "chance" is only a word invented by humans to conceal our ignorance. It explains nothing. If we perfectly understood all the laws of motion, we could infallibly predict whether a coin will come down heads or tails. A Christian believes that God *does* perfectly understand His own laws and knows which side up the coin will land, but Epicureans and neo-Darwinians believe that *nobody* knows!

All things bright and beautiful,
All creatures great and small,
All things wise and wonderful—
The Lord *Chance* made them all!

Do we want this taught to our children? (Watson 1976, 98–99)

Discussions such as these prepare the way for some of the most global objections to evolutionary theory.

In a peculiarly uncomprehending fashion, Watson alludes to an important distinction. When we describe something as a random process or an event as occurring by chance, there are two very different things that we may have in mind. Think of the processes that we often describe as "random processes," which are used in "games of chance": tossing coins, throwing dice, spinning wheels, shuffling and dealing cards. In all of these cases, we regard the final outcome of the process — the coin landing heads, the dealing of a full house or a lay down six spades, or whatever — as fixed in advance by the initial state of the system. Given the position of the coin on the gambler's thumb, the original impulse determined its trajectory. From statements describing those initial conditions, together with the laws of mechanics, it would be possible to deduce a statement describing that trajectory. Similar considerations apply to the throw of the dice, the spinning of the roulette wheel, and the dealing of the cards. I shall describe processes like these by saying that they are *apparently random*, but have a *deterministic basis*. In each case the deterministic basis is the full description of the beginning state of the system together with the laws needed for deducing a description of the outcome. The process is apparently random because we are ignorant of at least part of the deterministic basis. We use words like *chance* and *random* to indicate our ignorance.

We should separate apparently random processes from *irreducibly random* processes. An irreducibly random process is one that has no deterministic basis. That is, for an irreducibly random process, there is no set of laws of nature that can be applied to a complete description of the initial state of the system to permit the deduction of a description of the outcome. Or, to put the point simply, the way the process begins does not fix its ending. It is natural to wonder whether there are any irreducibly random processes. A long tradition of scientists and philosophers had maintained that the appeal to chance must always be a signal of our ignorance. Perhaps some processes *appear* to us to be random, but a more perfect view of their workings would reveal a deterministic basis. In Einstein's famous phrase, God does not play dice with the universe.

Contemporary physics, specifically quantum mechanics, inspired this rebuke from Einstein when it questioned the tradition. From the perspective of quantum mechanics there are indeed irreducibly random processes. A well-known example is the radioactive decay of atomic nuclei. Quantum mechanics claims that a complete description of the initial state of a nucleus, together with the laws of nature, does not deductively imply a description of the subsequent process of decay.

The most that can be derived from the statement of initial conditions is a description of the *probabilities* of outcomes in various ranges. Oversimplifying a bit, quantum mechanics does allow us to answer the following question: Given a collection of atomic nuclei in a specified initial state, what is the expected time for half the nuclei to decay to a specified final state? It is important to recognize that, in maintaining that irreducibly random processes exist, contemporary physics does not propose that those processes are lawless or unordered. Instead, it is claimed that the fundamental laws of physics are probabilistic. A *probabilistic* law is a statement asserting that, in a particular type of situation, a particular type of outcome will occur with a particular probability. An example of a probabilistic law is the claim that specifies the probability that atoms of a particular element will decay to form a certain product within a particular time interval. *Deterministic* laws, by contrast, assert that, whenever certain conditions are met, a particular type of outcome *inevitably* occurs. (I should point out that a small minority of physicists continues to hope that the processes currently regarded as irreducibly random will turn out to have a deterministic basis. The majority of physicists accept the existence of irreducibly random processes because certain technical theorems appear to show that the assumption of a deterministic basis would contradict experimental findings.)

The crucial question is, What do evolutionary theorists mean when they talk about mutation, for example, as a random process? The answer is complicated. In some cases, mutation is only *apparently* random. Mutations are often produced by modifying the genetic material in accordance with (nonprobabilistic) laws of physics and chemistry. Given a sufficiently complete description of a chromosome and its physicochemical environment, we could apply the laws of physics and chemistry to deduce a description of the subsequent state of the chromosome. We could deduce that certain bonds will be broken and that a nucleotide will be altered. In such cases, the mutation would be revealed as having a deterministic basis.

Other examples of mutation may be irreducibly random precisely because considerations from quantum physics become significant. Mutations are often produced through x-ray irradiation, and a full description of such episodes will require use of the probabilistic laws of atomic physics. Irreducible randomness enters the picture only when the details of the behavior of individual protons and electrons become important in effecting changes in the genetic material. Moreover, even the irreducibly random processes of mutation are not sheer chaos.

They are governed by definite physical laws: the probabilistic laws of quantum physics.

When evolutionary theorists claim that mutations are random processes, they intend to indicate that, in some cases, the deterministic basis of mutation is unknown, and that, in other cases, mutations are irreducibly random phenomena governed by the probabilistic laws of quantum physics. But they also intend something quite different: to exclude a certain type of basis for the occurrence of mutations.

Lamarck claimed that variations arise as they are needed. Darwinian evolutionary theory disagrees, emphasizing that mutations in a species do not arise in response to the requirements of the species. In other words, even though it might be advantageous for members of a species at a particular time to develop a particular mutation, that fact has no bearing on the mutations that actually occur. These points are sometimes made lucidly and explicitly in presentations of evolutionary theory: "To sharpen the contrast with Lamarckian ideas of the environmental induction of evolutionary changes, evolutionists stress the randomness of mutations. Since this term has often been misunderstood, it must be emphasized that it merely means (a) that the locus of the next mutation cannot be predicted and (b) that there is no known correlation between a particular set of environmental conditions and the particular allele among many possible ones to which a gene will mutate" (Mayr 1970, 102).

Our inability to predict the locus of a mutation may result simply from our ignorance of the precise physicochemical state of the chromosome and its environment. Or it may derive from the fact that events at the subatomic level are crucially relevant, so that even with perfect knowledge we could only assign probabilities to mutation events at individual loci. Molecular biology has made great strides in identifying the chemical changes that are produced in various kinds of mutations. It is now understood, for example, why some loci are much more likely to give rise to mutations than others. While processes of mutation are at least apparently random, and some of them may be irreducibly random, they are all subject to the laws of physics and chemistry.

Creationists standardly make two mistakes. They assimilate *apparent randomness* to *irreducible randomness*, and they overlook the fact that processes that are irreducibly random may be governed by probabilistic laws. Thus Wysong confuses irreducible randomness with sheer chaos. The fact that subatomic phenomena conform to probabilistic rather than deterministic principles does not prevent the possibility of general regularities in the behavior of large ensembles of atoms. Quantum chemistry explains in great detail why the elements regularly combine in just the ways that they do. Thus there is no reason to wonder whether the order found in biochemical reactions proceeds from some mysterious source.

By contrast, Watson and Morris confound the difference between apparent randomness and irreducible randomness. Contrary to what Morris may think, evolutionary theorists are not committed to the view that DNA was produced by an irreducibly random process. Of course, they do deny that there is any *goal-directed* process leading to the formation of DNA — as if, like Emerson's worm, the molecules in the primeval soup aspired to higher things, ascending through all the spires of form! But that is perfectly consistent with supposing that the DNA formed when a system of less complicated molecules (about whose relative abundances we are ignorant) underwent chemical combination according to the general laws that govern chemical reactions. There is no reason to believe *that* to be impossible. Morris's remarks, quoted above, trade on a rhetorical device, the device of confusing notions of randomness.

As I noted earlier, Watson seems dimly aware that there is a difference between apparent and irreducible randomness. However, he attributes to evolutionary theory the idea that even coin flipping is an irreducibly random process. Taking the further step of confusing irreducibly random processes with chaotic processes, ungoverned even by probabilistic laws, he produces his parody of a famous hymn. As should be evident, if evolutionary theorists wanted to expound their doctrine in verse, Watson's stanza would not be a candidate.

I have dealt with this topic at the beginning because Creationist objections are so free in using (or misusing) the idea of randomness. Let me now take up those objections, beginning with one that is both global and extraordinarily common in Creationist writings.

Entrapped by entropy?

Creationists love the laws of thermodynamics. Certainly, these laws are worth celebrating as major achievements of classical physics. Yet we might suspect the rhapsodies in which Creationists indulge: "It is well to note at this point, the implications of the First and Second Laws of Thermodynamics with respect to the origin of the universe. It should be stressed that these two Laws are *proven* scientific laws, if there is such a thing" (Morris 1974a, 25). Morris and his fellows are not moved to applause by their fine appreciation of thermodynamics and the evidence that corroborates it. Their accolades are intended to tell us which side to choose when we find that evolutionary theory and thermodynamics are in conflict.

The first law of thermodynamics is a principle about energy conservation. In its original form, it asserts that mechanical energy is equivalent to energy in the form of heat. I shall not be concerned here with a more precise statement of the law, nor shall I investigate the arguments, which occasionally appear, suggesting that the first law contradicts evolutionary theory. The reason is that, much as they love the first law, Creationists are even more devoted to the second.

A key concept in the second law of thermodynamics is the concept of entropy. There are various ways to understand entropy. One approach is to take the entropy of a system to be a function of that energy in the system that is unavailable for work. An ordinary engine, such as a steam engine or an internal combustion engine, can exploit only some of the energy stored within it. Some of the stored energy is locked away and unusable. To compute the entropy of the engine is to figure out how much of its internal energy is locked up in this way. The metaphor of locking up energy is easily understood by example. If we have two bodies at different temperatures, the heat energy of the colder body will not spontaneously flow to the warmer, thereby increasing the temperature difference; in this sense the heat energy of the cooler is "locked up." (An alternative way of viewing the entropy of a system is as a measure of the disorder of the system. In addition, the entropy concept can be introduced in terms of information theory. I shall follow the approach of classical thermodynamics, in which entropy is seen as a function of unusable energy. But the points I make will not be affected by this choice.)

The second law of thermodynamics is concerned with *closed* systems. From the thermodynamical view, a system is open if it is exchanging energy with what lies beyond its boundary. Closed systems are those that neither give energy to their environment nor receive energy from it. We may imagine a perfectly insulated box containing objects of various temperatures. Energy is exchanged among the objects in the box; some of them warm up, others cool down. (The hotter objects cool down, the colder warm up, until an equilibrium is reached.) However, there is no energy flow across the surface of the box. This imagined box would constitute a thermodynamically closed system.

We can now state the second law: The entropy of a closed system increases with time. What this means is that if we have a thermodynamically closed system, like our imagined box, the total amount of energy within it remains constant through time, but an increasing proportion of that energy becomes unavailable to do work. (Alternatively, the system becomes even more disordered.) The formulation that I have given accords with those found in textbooks on physics (for example, Morse 1974). But it does not coincide with the statements of the second law offered by some Creationists.

Creationists like to present the second law either by omitting any mention of its restriction to closed systems or by choosing a statement that does not make this restriction clear. Consider the following passages:

The second law of thermodynamics is the law of increasing entropy, stating that all real processes tend to go towards a state of higher probability, which means greater disorder. This law applies to all known systems, both physical and biological, a fact which is universally accepted by scientists in every field. (Morris 1974b, 98)

Basically the second law says three things:
1. Systems will tend toward the most probable state.
2. Systems will tend toward the most random state.
3. Systems will increase entropy, where entropy is a measure of the availability of energy to do useful work. (Wysong 1976, 241)

The Second Law of Thermodynamics, also universal and invariable beyond any scientific doubt, states that, although the total amount of energy remains unchanged, in all real processes, some of the energy involved (all processes in the universe, be they physical, geological, biological, etc. involve energy transformations) is transformed into non-reversible heat energy and is no longer available for work. (Hiebert 1979, 108–109)

In all these cases, Creationists fail to acknowledge that the second law states only that the entropy of *closed* systems increases. Elsewhere, they quote a statement by Isaac Asimov, intended to provide a non-technical explanation of the second law: " 'In any physical change that takes place by itself the entropy always increases.' (Entropy is 'a measure of the quantity of energy *not* capable of conversion into work.')" (Morris 1974a, 38–39; quotes are from Asimov 1970, 8). Asimov's statement is not inaccurate. But its reference to closed systems—borne by the phrase "takes place by itself"—is easily overlooked.

Why is the second law supposed to constitute a problem for evo-lutionary theory? The basic idea is simple. According to the second law, "There is an inexorable downhill trend toward ultimate complete randomness, utter meaninglessness, and absolute stillness" (Morris 1974b, 121). However, *any* Darwinian history presupposes an "upward trend" toward increasing complexity and organization. Faced with the "anti-evolutionary implications of the Second Law of Thermodynamics," evolutionary theorists are supposed to trot out a number of standard desperate responses: The second law "does not apply to living systems"; The law is "a statistical statement" with possible exceptions; The law may not always have operated; and so forth (Morris 1974a, 40–42; Morris 1974b, 121–122). As an afterthought, evolutionary theory is allowed its most obvious response: The second law applies only to closed systems.

Let us be completely clear about the logic of the situation. Evolutionary theory would contradict the second law if (and *only* if) the construction of Darwinian histories required us to suppose the existence of thermodynamically closed systems in which entropy does not increase. But no such supposition is required. Darwinian histories do presuppose that large amounts of energy remain available for work in large numbers of systems of living things. Let us consider the following kinds of systems: an organism, a genealogy including an initial pair of organisms and their descendants, a lineage consisting of an ancestral population and its descendants, and the total lineage that comprises all life on earth. In many systems of these kinds, entropy *decreases* over time. But the systems in question are all open. Energy from the sun is constantly entering the system comprising the earth and its inhabitants. That energy flows into lesser organic systems as they feed and warm themselves.

Creationists are very vague about exactly where evolutionary theory contradicts the second law. They do not point to some particular feature of Darwinian histories and show that that feature presupposes a system that thermodynamics disallows. However, it may help to dissolve any lingering worries if we look at life on earth from the perspective of classical thermodynamics. Since the earth receives large amounts of energy from the sun, we cannot think of the earth as a closed system. By broadening our horizons, however, we can view the earth as part of a closed system. That closed system will include the earth, the sun, and those regions of the universe that exchange energy with them. (It will be extremely large.) Classical thermodynamics tells us that, within this vast closed system, entropy increases. It says absolutely nothing about entropy variation at the local level. Thus the fact that the system contains pockets in which entropy decreases (for example, the subsystem comprising terrestrial life) is perfectly compatible with the laws of classical thermodynamics.

Unfortunately, this is not the end of the story. Although they do not refer to closed systems in stating the second law, Creationists have heard that the law only applies to such systems. So they are ready for the response I have just given. Morris even calls it "an exceedingly naive argument" (Morris 1974b, 123). There are two popular Creationist rejoinders. The first is to pooh-pooh the concept of a closed system. The second is to change the subject.

There is something curious about the first rejoinder. Here are two passages from Morris:

Although it is true that the two laws of thermodynamics are defined in terms of isolated systems, it is also true that in the real world there is no such thing as an isolated system. *All* systems in reality are open systems and, furthermore, they are all open in greater or lesser degree, directly or indirectly, to the energy from the sun. Therefore, to say that the earth is a system open to the sun's energy does not explain anything, since the same statement is true for every other system as well! (Morris 1974, 43)

Obviously growth cannot occur in a closed system; the Second Law is in fact *defined* in terms of a closed system. However, this criterion is really redundant, because in the real world closed systems do not even exist! It is obvious that the Laws of Thermodynamics apply to open systems as well, since they have only been tested and proved on open systems! (Morris 1974b 125)

For all the laurels heaped on classical thermodynamics, it appears that physicists have been somewhat misguided. They have introduced an unnecessary restriction, so that the laws of thermodynamics are not applicable to any existing systems. How fortunate that the Creationists are able to set them straight.

The concept of a thermodynamically closed system, like that of a frictionless plane or a perfectly rigid body, is an *idealization*. Let us recall our imaginary example of a closed system. I envisaged a perfectly insulated box within which bodies exchanged energy. Reality contains no such boxes. What we find are approximations to perfect insulation. To put the point more exactly, we find systems that can be treated *as if they were closed* because their energy exchange with the external environment is negligible in comparison with the energy flow within them. The laws of thermodynamics can be tested and confirmed (*not* proved) by investigating such systems, just as the laws of rigid-body mechanics can be tested and confirmed by investigating the behavior of things (like objects made of wood or metal) that are approximately rigid. What we cannot do is apply the laws of thermodynamics to systems for which the energy flow across the boundaries is significant in comparison with the energetic transactions within them. That would be like trying to use rigid-body mechanics to explain the motion of blobs of jelly. Because the systems studied in evolutionary theory exchange vast quantities of energy with the environment, they cannot be treated as if they were closed.

The Creationist response, "Closed. Open. It's all the same," displays a hopeless misunderstanding of the second law of thermodynamics, born of ignorance of the role of idealization in physics. The second rejoinder appears to be more sophisticated. Morris advances what I shall call the "evolving junkyard" argument:

It should be self-evident that the mere existence of an open system of some kind, with access to the sun's energy, does not of itself generate growth. The sun's energy may bathe the site of an automobile junk yard for a million years, but it will never cause the rusted, broken parts to grow together again into a functioning automobile. A beaker containing a fluid mixture of hydrochloric acid, water, salt, or any other combination of chemicals, may lie exposed to the sun for endless years, but the chemicals will never combine into a living bacterium or any other self-replicating organism. More likely, it would destroy any organism which might accidentally have been caught in it. Availability of energy (by the First Law of Thermodynamics) has in itself no mechanism for thwarting the basic decay principle enunciated by the Second Law of Thermodynamics. *Quantity* of energy is not the question, but *quality*! (Morris 1974b, 123)

As the history of science teaches us, great new ideas are seldom the property of a single individual. Wysong achieves the same insight: "If the decreased entropy and high orderliness of life is accounted for solely on the basis of open system thermodynamics, you might ask why other open systems don't likewise experience such ordering? [sic] In other words, why don't battered Volkswagons [sic] in junkyards order themselves into shiny new Cadillacs? A junkyard is an open system" (Wysong 1976, 244).

These passages might deceive us into thinking that a clever objection has been launched. But the issue has been shifted. Evolutionary theory was originally challenged to reconcile its claims of increasing organization and complexity with the second law of thermodynamics. The challenge is met by pointing out that there is no contradiction because the systems that are supposed to generate the trouble cannot be treated as good approximations to the ideal of a closed system. Creationists now ask why some open subsystems show decreasing entropy and others (cars in junkyards) do not. That is an entirely different question, and one that has an obvious answer. The simple answer is that the open systems that do not evolve have a different physiocochemical makeup from those that do evolve. Steel is significantly different from

DNA. A detailed answer would explain exactly how both living things and automobiles change, in different ways, in accord with the laws of physics and chemistry. Even though the two systems are relevantly similar from a thermodynamical point of view, their physicochemical make-ups are relevantly different. Contrary to Wysong's suggestion, evolutionary biologists do not suppose that open-system thermodynamics *accounts for* the decreased entropy of life. Nobody alleged that having an open system *is sufficient* for decreased entropy. The point was to rebut a charge based on ignoring the restriction of the second law to closed systems. Evolutionary theory contends that decreased entropy is *possible* in an open system, not that it *must* happen in *any* open system.

To see how bad this argument really is, let me construct an analogous line of reasoning. J. Fred Hailey, the attorney for the defense, offers an apparently conclusive case for his client's innocence. The victim could not have been done in by the defendant because the fire in the grate had consumed all the oxygen, so that the unfortunate party was dead prior to the defendant's arrival. Alas, the prosecutor has an eye for details and points out that a window was open, so that plenty of oxygen would have been available to keep the victim alive. But the defense never rests. Hailey retorts: "Window, no window, what's the difference? Besides, if the oxygen could keep the victim alive, why couldn't it revive the sofa on which he was sitting? Answer that, if you can!" The prosecutor is left (virtually) speechless.

Lurking behind the Creationist argument we find, once again, the randomness ploy. Morris and Wysong both use the "evolving junkyard" objection to conclude that some mechanism must be present in those open systems where entropy does decrease. They regard this conclusion as evidence that a designing hand guides the course of life. But this is a gigantic *nonsequitur*. Evolutionary theory never assumes that there is no explanation for the flow of energy in systems of living things. On the contrary, it supposes that physicochemical changes occur, and that energy is exchanged, in accordance with precise physical and chemical laws. Nonetheless, Creationists encumber the theory with the assumption that random processes are supposed to give rise to order and complexity. Wysong writes, "All observations confirm the inability of randomness to transform itself (open or closed system, it makes no difference) into high order" (Wysong 1976, 252). Such remarks thrive on confusing apparent randomness with irreducible randomness, and irreducible randomness with sheer chaos. According to evolutionary theory, the physicochemical constitution of certain open systems, together with the physicochemical features of their environment, causes those systems to maintain or decrease their entropy. Other open systems, with different constitutions or different environments are not so lucky. Classical thermodynamics does not preclude the possibility of either type of system because classical thermodynamics concerns itself with closed systems.

I would not have dealt with this objection at such length if the appeal to thermodynamics were not so popular a device among Creationists. (It is a mainstay of Creationist literature, and was duly discussed at the Arkansas trial.) However, my discussion does expose an interesting set of Creationist tactics. The original suggestion of contradiction between evolutionary theory and thermodynamics thrives on misstating the second law. Next, the Creationists consider responses that do not correct the formulation. This enables them to portray evolutionists as invoking the concept of a closed system out of desperation. They continue by distorting the role of idealization in science, and by changing the question. Instead of explaining how the evolution of systems of living things is consistent with the second law, scientists are supposed to *use* the second law to explain the precise details of energy exchange. Enter the randomness ploy, and the net result is confusion.

Mutations and misfits

When biologists construct a Darwinian history, they advance claims about the emergence of advantageous variations in populations. Population genetics tells us that there are two sources of variation. Novel genotypes may arise either through recombination or mutation. Of these two sources, mutation is the more fundamental. For mutation is the "ultimate source of genetic novelties" (Mayr 1970, 98). Because evolutionary theory attempts to understand the spread of characteristics by reference to the advantages they confer, it presupposes that there are mutations that produce genes for beneficial characteristics.

So far, we have a straightforward, if oversimplified, account of the commitment of evolutionary theory to a genetic thesis. The Creationists believe they can show that the thesis is false. With his usual zeal, Hiebert reveals that there has been a cover-up. The sources of inheritable change include "MUTATIONS, which are genetic injuries, or mistakes, and, to be candid, *always affect* the viability (ability to compete and survive) of the organism *adversely*. This cold, hard fact regarding mutations is rarely discussed, or even admitted publicly by evolutionists" (Hiebert 1979, 37). Morris presents the attack in more detail. He claims, in succession, that mutations are random, uncommon, and not usually beneficial. As we shall see, there is some truth in the vicinity of each of these points. However, Morris cannot resist going for the jugular: "As a matter of fact, the phenomenon of a truly beneficial mutation, one which is *known* to be a mutation and not merely a latent characteristic present in the genetic material but lacking previous opportunity for expression, and one which is permanently beneficial in the natural environment, has yet to be documented" (Morris 1974a, 56). Hence Morris, like Hiebert, concludes that "the net effect of all mutations is harmful."

Purged of rhetorical excesses, the criticism is that the theory and experimental study of mutation conflict with a presupposition of all Darwinian histories. Two allegations of inconsistency are relevant: Mutations are rare and harmful; Darwinian histories suppose them to be common and beneficial. One of these points is relatively easy to disarm. As the eminent authors of an excellent textbook put it, "Mutations are either rare or ubiquitous events depending on how we choose to look at them. The mutation rates of individual genes are low, but each organism has many genes, and populations consist of many individuals" (Dobzhansky, Ayala, Stebbins, and Valentine 1977, 71). This statement rests on experimental evidence that allows scientists to compute an average rate of mutation per locus for organisms of a given species. Estimation of the number of loci in the genome generates a figure for the number of mutations per zygote (fertilized egg). Multiplying this figure by the number of individuals constituting one generation of the population, one obtains a projection of the number of mutations arising in one generation. For *Homo sapiens*, the authors take this number, in the current population, to be about 8 billion.

The charge that mutations are rare depends on confusing the mutation rate per locus (of the order of 1 mutation per 100,000 loci) with the rate per zygote (of the order of 1 mutation per zygote) or the rate per population (of the order of 1 billion per population). From an evolutionary perspective, it is the last of these rates that is important. Hence, although Morris is right to claim that mutations are rare (in *one* sense), he is quite wrong to think that this spells trouble for evolutionary theory. Indeed, neo-Darwinian evolutionary theory insists on the rarity of mutation at any individual locus, claiming, for this reason, that natural selection is a more powerful evolutionary force than mutation. (If mutations were extremely frequent, then selection would play a less crucial role.)

The point about the harmful effects of mutations is trickier. A mutation is an internal change in the genetic material (see chapter 1).

Such changes come about through deletion of bases, insertion of bases, or the substitution of one base for another. If a mutation occurs in a structural gene (a gene that codes for a polypeptide chain), then there is a range of possible effects. Substitution can produce a gene that directs production of the same polypeptide, of a polypeptide that is different but functionally equivalent, or of a polypeptide that is incapable of performing the same function. The polypeptide formed depends on the sequence of bases in the segment of genetic material. However, different sequences can carry the same message; the "genetic code" is redundant. Hence it is clear that some mutations will leave the message intact, and thus lead to production of the same polypeptide. Study of the genetics of human hemoglobin discloses the other possibilities. Some mutations, which are presumed to arise from base substitution because they involve change of only one amino acid in the polypeptide chain, generate polypeptides that are functionally equivalent to the normal product. Unfortunately, others, like the sickle-cell mutation, do not.

This very simple outline of some of the ways in which mutations can occur neglects some aspects of mutation that may be extremely important for studies of evolution. For example, I have not discussed the ways in which mutation might affect those genes involved in regulatory functions. However, this much will suffice for present purposes. Let us first ask whether the account I have given allows for the possibility of beneficial mutations.

The first point to make is that mutations are harmful, neutral, or advantageous *relative to a particular environment*. Although the effects of some mutations are so severe that they prove lethal in a wide range of environments, evolutionary theory is not committed to the idea that there are mutations that prove beneficial in all, or even most, environments. Evolutionists frequently assume that there were mutations that were advantageous to members of an ancestral population in their particular environment. But the environment includes not only the inorganic world, organisms of other species, and conspecific organisms. The benefits conferred by a mutation also *depend on the genetic background in which it is set*. No gene is an island. As Mayr puts it, "The selective value of a gene, its 'goodness' is determined by a complex constellation of factors in the external and internal environment. . . . A gene may add to viability on one genetic background but have the characteristics of a lethal on another genetic background" (Mayr 1970, 101). For example, a base substitution may cause production of a slightly abnormal protein. In combination with the rest of the genotype, production of that protein may prove advantageous, given the external environment. In other environments or against other genetic backgrounds, the effect of the abnormal protein might be disastrous.

Now there are many experimental demonstrations of the possibility of adaptive mutations arising in special strains of organisms placed in laboratory environments. The genetics of some organisms—specifically some viruses, some bacteria, and some molds—is well understood at the biochemical level. Biologists are able to develop strains of these organisms that are unable to produce some protein necessary for their survival in a particular environment. They can then show that mutants who do not have the deficiency do arise. However, this is not enough to satisfy Morris. What Morris demands is a mutation that is "permanently beneficial in the natural environment." I am not sure what force is to be given to the word "permanently." With possibly misplaced charity, I shall assume that Morris is not making the mistake of requiring a mutation that is advantageous in *all* natural environments. As I interpret him, he is dissatisfied with laboratory demonstrations of the kinds I have mentioned because in those demonstrations scientists have concocted peculiar strains of organisms and placed them in peculiar environments. He would like something more natural.

We can accommodate him. There are numerous examples of advantageous mutations arising in natural populations. One well-known family of cases is the development in many species of insects of resistance to pesticides (notably DDT). Another type of case includes many examples of plant tolerance of metals in the soil near mines. Why does Morris find these examples unconvincing?

There are two possible reasons, neither of which amounts to very much. One is that all of the instances I have mentioned involved adaptation to an environment that man has modified. This is spurious. Each of the cases shows just what evolutionary theorists would expect: Although natural populations are well adapted to their environments, alteration of their environments can make mutations advantageous. The fact that humans have done the altering is irrelevant. Any natural process that produced the same, or analogous, environmental effects would have generated the same, or analogous, consequences.

The second reason is that we do not really *know* that the variations in question are due to mutations. Perhaps it may turn out that the variations are really the result of recombination, or of "a latent characteristic present in the genetic material but lacking previous opportunity for expression." This response gets high marks for low cunning. How can we discover that natural populations of houseflies (for example) acquired mutant alleles for resistance to DDT? But skeptical worries are extravagant. The biochemical basis of some examples of pesticide resistance (in insects and in rats) is now understood; a change at a single locus can produce (or increase the rate of production of) enzymes that enable the pesticide to be detoxified. Biochemical studies have shown that there are very close structural similarities among enzymes that confer different degrees of resistance, thereby providing excellent evidence for the conclusion that alleles for high resistance originally arose by mutation. (It is important to note that the mutations in question may well have occurred long before the insect population encountered the pesticide. Darwinian evolutionary theory does not accept the Lamarckian idea that environmental needs induce mutations.)

There is no reason to doubt that mutations that are advantageous in natural environments sometimes occur. Reflection on the molecular biology of the gene teaches us that this is what we ought to expect. There is no denying that mutations in structural genes (those that direct the formation of polypeptides) are most likely to prove either neutral or deleterious. Deletions and additions will probably create situations in which necessary proteins are not available. Some substitutions will produce nonfunctional proteins; others will generate a functionally equivalent protein or even the normal protein itself. But *molecular biology recognizes a further possibility*. Substitutions can lead to the production of proteins that are able, in combination with the rest of the gene products, *better* to perform the functions of normal proteins or to perform new and useful tasks. Experimental results show us that this possibility is sometimes realized. Evolutionary theory can have the mutations it needs.

Trying time

Ever since 1861, when Lord Kelvin argued that there has been too little time for the evolution of contemporary organisms, critics of evolutionary theory have had a field day with this type of criticism. Kelvin's original argument was ingenious and precise. Observation of living organisms, together with historical records of past organisms, can enable us to figure out the fastest possible rate at which large-scale evolution occurs. Domestic animals have not evolved since the pharaohs (whose people obligingly left good drawings), so we have some gauge on the time it takes for significant evolution to occur. Nothing much has happened in 4,000 years. From the other direction, geophysical estimates of the age of the earth, together with an evolutionary proposal about the development of life, enables us to figure out the slowest rate at which evolution could have occurred. For evolutionary change cannot go at so slow a pace that the sequence of changes would occupy more than the allotted span. Now what Kelvin argued was precisely that the slowest rate at which evolution could get the job done, given the available time, is *faster* than the fastest rate at which evolution can work. So an evolutionary account of the entire development of life cannot be correct.

We now know that Kelvin was wrong. Although his concerns were entirely justified, his argument rested on a mistake about the earth's age. Quite understandably, Kelvin knew nothing about the energy due to radioactive decay. His estimate was based on two procedures. One applies the theory of heat diffusion to the earth, *assuming that no heat is generated within the earth's surface*. The other applies a classical model of the source of the sun's heat to compute the time for which the sun could have "illuminated the earth." Once the phenomena of radioactive decay were discovered, it was clear that Kelvin's estimates of the age of the earth are too small. For radioactive decay is a source of heat energy within the earth, and a source of heat within the sun; neither of these sources appear in Kelvin's calculations.

Like other early objections to Darwinism, Lord Kelvin's argument has been dusted off and refurbished by the Creationists. Two Creationists have recently tried to correct the faulty estimate of the earth's age, arguing that, even when radioactive decay is taken into account, there is too little time for evolution to have occurred (Slusher and Gamwell 1978). H. S. Slusher and T. P. Gamwell proceed by reconstructing Fourier's theory of heat diffusion for several special cases (sphere with no heat source and semiinfinite solids with various types of heat source). Now one might wonder what relevance any of these cases has to the problem of the age of the earth. After all, although the earth is (roughly) a sphere it is not a sphere without a heat source — that was the unforeseen error in Kelvin's calculations. Certainly it is not a semiinfinite solid. Slusher and Gamwell try to justify treating the earth as a semiinfinite solid by pointing out that, in the case where no heat source is present, approximately the same results are obtained by modeling the earth as a semiinfinite solid or by treating it as a sphere (Slusher and Gamwell 1978, 28). But this argument is specious. The differential equation to be solved when a heat source is present is not the same as the differential equation to be solved when no heat source is present. The fact that we obtain the same results from solving the latter equation for the case of a sphere and for the case of a semiinfinite solid does not mean that we shall get the same results if we solve a quite different equation for both cases. Slusher and Gamwell fail to show that a realistic model of heat diffusion within the earth generates a short timescale.

Far more prominent in Creationist writings is a different kind of argument for the claim that there is too little time for evolution. This reasoning, which turns on combinatorial considerations, is the brainchild of the tireless Henry Morris. It is featured in the opening chapter of one of his works (Morris 1972), is lauded by a fellow Creationist (Hiebert 1979, 98), and achieves stardom in another of his works (Morris 1974a). Here is one version. We are going to estimate the "probability of increasing complexity of living systems": "The problem is how can a population of living organisms structured at one degree of complexity be elevated by random processes to a higher degree of complexity? [sic]" (Morris 1974a, 66). Neo-Darwinian evolutionary theory assumes that the "elevation" will be accomplished by small discrete steps. The first task is to estimate the probability of each step: "Again, however, let us be as generous as possible, and assume that each successive evolutionary step has a probability of success of $1/2$. That is, a given population representing, say n degrees of order (information content in its genetic code) has as great a probability of changing to a population of $(n + 1)$ degrees of order as it does of slipping back to $(n - 1)$ degrees of order or lower" (Morris 1974a, 67). Morris continues by pointing out how generous he has been. Since harmful mutations occur more frequently than beneficial mutations, the population is likely to "slip backward." Waiving this point, he proceeds to compute the probability of the "evolution of higher organisms": "For one kind of animal to evolve into a distinctly higher kind of animal would require a tremendous number of mutational steps. Huxley's example, previously quoted, mentioned a million mutational steps for the assumed evolution of a horse. Considering that mutations must be small, each one probably imperceptible, a million seems small indeed" (Morris 1974a, 68–69). Now it is possible to apply an elementary theorem of probability theory. Suppose that we are to make 1,000,000 independent trials in sequence, and on each trial the probability of success is 1 in 2; then the probability of success on every trial is 1 in $2^{1,000,000}$. (The probability of 2 heads in a sequence of 2 coin tosses is $1/4 = 1$ in 2^2, 3 heads in 3 coin tosses is $1/8 = 1$ in 2^3, and so forth.) $2^{1,000,000}$ is approximately $10^{300,000}$, so the probability of complete success through 1,000,000 trials is obviously very small. Here is Morris's triumphant conclusion:

The universe of 5-billion light-years radius contains only 10^{80} particles of electron size. If there were no empty space at all, with the entire universe solid-packed with electrons, it could still hold only 10^{130} electrons. If each such electron were a mutating system, going through the required million mutations a billion times every second for the 10^{18} seconds in 30 billion years, the total number of attempts that could be made is only 10^{157}. There is not the remotest possibility that one of these would be successful, since the chance of one success is only one out of $10^{(300,000-157)}$ or one out of $10^{299,843}$. (Morris 1974a, 69)

Because of its apparent numerical precision, Morris's argument may sound very convincing. Moreover, Morris has actually computed *something* correctly. He has shown that if we define a *grand success* as a sequence of a million independent trials, each of which is successful, and if we have a probability of 1 in 2 of being successful on each trial, and if we have 10^{157} opportunities for a grand success, then there is a very small chance of achieving a grand success. The crucial question concerns the relevance of this result to the theory of evolution. Does evolutionary theory presuppose the existence of grand successes?

Let us begin by asking what exactly we are supposed to be computing. The natural way of interpreting Morris is as follows. We imagine a population of very primitive organisms and compare them with a population of contemporary organisms, such as horses. We estimate that they differ by a million small gene modifications. Now we order those gene modifications to form a sequence through which the horse could have evolved. Evolutionary theory assumes the existence of a sequence like this, and is thus committed to the existence of a grand success.

One point that should be recognized immediately is that the randomness ploy is rampant here. Frequently, there are ways of under-describing the starting conditions of an actual occurrence so as to make it look extremely improbable. Imagine that you are dealt a hand of thirteen cards, one after the other, from a standard deck of fifty-two cards. In Morrisian style, we might ask for the probability that you would have been dealt those cards in exactly that order. The answer is about 1 in 4×10^{21}. Of course, that makes the event look very improbable. But it happened.

Cases like these should mystify us only if we overlook the obvious fact that the probability of an event can be very low relative to a *particular description of its initial conditions* even though those initial conditions actually determine that the event will occur. When you are told only that you would receive thirteen cards from a standard deck, being dealt that particular sequence of cards seems very improbable. However, the sequence is completely determined by the conditions prior to the deal—specifically, by the way in which the cards are arranged in the pile. Similar remarks apply to the evolution of the horses. When one is told only that a particular sequence of mutations is needed to produce horses from primitive organisms, the odds look fantastic. Yet that is perfectly compatible with the fact that the initial conditions of the situation determine that that sequence will occur. Morris has confused apparent improbability with irreducible improbability, in exactly the same way that he and his fellow Creationists consistently conflate apparent and irreducible randomness. (It is interesting to note that, in a passage quoted by Morris, Julian Huxley makes some remarks about large-scale evolution that presuppose the point I have just made. Because he does not appreciate that point, Morris subjects Huxley to quite undeserved ridicule.)

However, there are more things amiss with Morris's attempt to connect his probability calculations with evolutionary theory. It is far from obvious what improbable sequence of events Morris thinks he can pin on evolutionary theory. Morris's remarks suggest that we should envisage a collection of primitive organisms. Each of these is allowed to play the *mutation game*: If it mutates "towards higher complexity" it stays in the game and is permitted to play in subsequent rounds; if it does not, it drops out; and if it persists through a million rounds, it becomes a horse. The argument seems to be that the chance of any one organism winning the jackpot (becoming a horse) is fantastically small, and that the initial collection of organisms is too small to balance the odds. This absurd idea would make Morris's probability calculations relevant to evolutionary theory by identifying some sequence of events to which the computation applies.

I am reluctant to assume that this scenario represents Morris's intentions simply because it is *so* ludicrous. Surely Morris is aware that organisms die. The trouble is that when we consider the way in which evolutionary theory would understand the evolution of horses, the type of probability calculation he performs appears irrelevant. From the evolutionary perspective, the initial state consists of a *population* of primitive organisms with a range of genotypes. The final state consists of a *population* of horses with a range of genotypes. The process leading from the initial state to the final state involves two different kinds of events: events of mutant *individuals* arising in the population and events of the spread of mutant genotypes through the *population*. Morris's computation, which seems to treat individuals of the population in isolation, ignores the second kind of event. This is a crucial lapse because, as a favorable mutant genotype spreads through the population under natural selection, the number of individuals in which further mutations can arise is increased. Furthermore, Morris seems to conceive of mutations as arising one at a time. Once we recognize the role of the population in evolution, it is quite clear that this is a mistake. Mutations are constantly occurring in populations. Many of them are eliminated quickly. But some spread through the population, and eventually become fixed. Of course, as they are spreading, further events of mutation occur. Because he fails to distinguish populations and individuals, Morris's calculations are meaningless. (As I have noted, Morris seems to conceive of the individuals of the population as persisting for extremely long times. I suspect that this arises from this failure to separate populations and individuals. In a sense, the population persists. But it is the individual organism that mutates. Morris's argument assumes that there is a single thing that is the seat of mutations and that persists, and it is this false assumption that generates the appearance of absurdity.)

Moreover, once we recognize that the issue should be addressed by considering the origin and spread of favorable mutations in populations, it is clear that the appropriate mutation rates to be considered are *population* rates of mutation per generation. Assuming a population of 1 million organisms, and standard mutation rates per zygote, a conservative estimate of the number of mutations per generation would be on the order of 100,000 mutations per generation. Waiving Morris's generous offer, let us suppose that 1 in 1,000 of these are advantageous in the population's environment. That allows for 100 advantageous mutations per generation. Given 25,000,000 generations, it is hardly implausible that 1,000,000 mutations would become fixed in the final population. (My calculation is highly schematic, but it is far closer to the actual processes envisaged by evolutionary theory than is Morris's computation. For more detailed calculations on shorter evolutionary sequences, see Simpson 1953, chapters iv and v, especially 109–110.)

Morris's other efforts at computation fare no better. In each case, we find the two mistakes that I have identified. (I ignore smaller errors such as Morris's odd ideas about populations "slipping backward" and his importation of an undefined idea of evolutionary progress.) The randomness ploy is invoked to blind the reader to the fact that evolutionary processes are governed by precise physicochemical laws. For example, the first chapter of the "popular exposition" of Creationism (Morris 1972)—a work that the Old Time Gospel Hour distributes *gratis*—moves quickly to a combinatorial argument. Morris introduces it as follows: "The idea that a complex structure or system can somehow be formed by chance is a persistent delusion accepted by evolutionists" (Morris 1972, 3). This is, at best, misleading. Evolutionary theory maintains that the initial state of the primeval soup determined subsequent formation of biologically significant molecules, that the initial state of a population and its environment determines its subsequent evolution, and so forth. Given certain partial descriptions of those initial states, the subsequent events may *appear* highly improbable. But, as we have seen, that phenomenon is both commonplace and unworrying.

The second error consists in performing probabilistic computations that are either inapplicable to evolutionary theory or applicable only by grossly distorting the theory. We have looked at one example of the error. Another, which is even more blatant, occurs in the opening combinatorial argument of the "popularization" (Morris 1972, 3–4). There, Morris takes evolutionary theory to be committed to the view that complex systems come into being through random shuffling of their parts. (For a lucid discussion of this particular argument, see Simon 1968, 90–95.) As he himself later admits (Morris 1972, 7), evolutionary theory is not committed to so peculiar an idea. However, his resolution of the issue consists in the argument whose credentials I have examined. It is hard to resist the impression that all these computations are designed to bamboozle those who become weak at the knees at the sight of numbers.

Fear of fossils

One of the most sustained, detailed, and apparently informed attacks on evolutionary theory is the appeal to the fossil record. In a book that has probably circulated more widely than any other piece of Creationist writing, Duane Gish articulately presents a series of variations on a common theme. The fossil record is incomplete. Because there are gaps in the fossil record, evolutionary theory is forced to make claims that can be shown to be highly implausible. Although this theme is sounded by other Creationists, Gish is its main exponent, and I shall address myself primarily to his presentations of it.

After an initial chapter raising the methodological questions about proof and falsifiability already discussed, Gish opens his attack by specifying what we ought to expect if life has indeed evolved: "If life arose from an inanimate world through a mechanistic, naturalistic, evolutionary process and then diversified, by a similar process via increasingly complex forms, into the millions of species that have existed and now exist, then the fossils actually found in the rocks should correspond to those predicted on the basis of such a process" (Gish 1979, 33). This sound innocuous enough. But trouble lurks in Gish's word *correspond*. The fossils actually found in the rocks should correspond to the past organisms described in Darwinian histories, in the sense that there should not be fossils that do not fit within the total account of the development of life. However, it would be wrong to think that *all* of the organisms described in Darwinian histories will have left mementos of themselves. Fossilization is far from being the inevitable destiny of a past organism.

As his argument proceeds, Gish sometimes makes it clear that he understands this point: "It is true that according to evolutionary geology only a tiny fraction of all plants and animals that have ever existed would have been preserved as fossils. It is also true that we have as yet uncovered only a small fraction of the fossils that are entombed in the rocks. We have, nevertheless, recovered a good representative number of the fossils that exist" (Gish 1979, 51). Although Gish seems to recognize the vicissitudes of fossilization, his remarks oversimplify the situation. The fossil record is not only partial. It is biased. Some organisms—such as marine invertebrates—are much more likely to be *preserved* as fossils than others. The fossil remains of some organisms—the teeth of large recent mammals, for example—are much more likely to be *found* than the fossil remains of others. (Paleontologists have realized in recent years that their predecessors failed to find the teeth of primitive mammals, all of whom were small, simply because of the sieving techniques that were used in collecting.) Hence Gish's optimistic assessment that the fossils we have represent the organisms that previously existed (Gish 1979, 50–51) is unjustified. There is a good fossil record for some groups of organisms and a bad fossil record for others. Moreover, there are independently testable hypotheses that explain why the fossil record is complete and incomplete in the ways that it is.

The opening chapter of a major contemporary text in paleontology gives a lucid exposition of this point. The authors warn, "Any study of fossils or use of paleontologic data must be based on a clear understanding of the strengths and weaknesses of the record" (Raup and Stanley 1978, 8). They continue by explaining the various ways in which dead organisms can be destroyed. Dead animals can be attacked by other organisms. They can be broken up by mechanical means (as, for example, when a skeleton falls apart or bones are smashed as debris is piled upon them). Or they can be decomposed chemically. Some fortunate organisms survive these hazards and leave traces of their presence. Mollusks, for example, have a better chance of being fossilized because they may quickly be buried and insulated against major forms of biological and mechanical destruction. Birds and insects, at the other extreme, are likely to die in places where sedimentation is not occurring and where their delicate skeletons will be extremely vulnerable. (For much more detail about the likelihood of fossilization and the experiments that have been performed to test claims about the modes of destruction, the reader should consult the excellent first chapter of Raup and Stanley 1978.)

Once the biased character of the fossil record is understood, we can see how Creationists can exploit the bias. First they can quote paleontologists who say that the fossil record is good. (For *in certain respects* it is.) Then they can point out that, for some evolutionary sequences, the fossil record does not exhibit the organisms that are supposed to have existed. The strategy is to ignore the good parts of the record and focus on groups of organisms that scientists know, *on independent grounds*, are unlikely to be fossilized.

Gish is not above using this strategy. So, for example, he lambasts evolutionary theory on the grounds that there is no detailed transitional series of fossils in any of the four independent cases of the evolution of flight (Gish 1979, 88–90). (Flight is standardly taken to have evolved independently in birds, bats, pterosaurs, which are extinct flying reptiles, and insects.) As we shall see later on, Gish's demands for transitional series have their own peculiarities. But, for the moment, the point that concerns me is the exploitation of bias in the fossil record. There is a clear and convincing explanation of why the fossil record for flying animals—especially insects—should be incomplete. Gish ignores this explanation and uses the bias of the fossil record to add rhetorical flourishes to his argument.

However, it should stand to his credit that Gish does try to analyze some of the examples in which paleontologists claim that the fossil record displays large-scale evolution. So he devotes several pages to discussions of some of the famous transitional series of fossils that are often taken to reveal the evolutionary links between classes of organisms. The three main examples are the links between fish and amphibians, reptiles and mammals, and reptiles and birds.

Let us begin with a brief outline of the views of contemporary paleontology. Amphibians evolved from a group of fishes called the *crossopterygian fishes*. These were bony fishes, related to lungfishes. The pattern of their skull bones is comparable to that found in early terrestrial vertebrates. Likewise, the paired fins attached to the pectoral and pelvic girdles show similarities to the limbs of early amphibians (Colbert 1980, 69–71). Fossils of early amphibians—the ichthyostegids—are different from fossils of crossopterygian fish in some respects, but there are numerous important common features. After noting the relations between the skulls of crossopterygian fishes and the skulls of ichthyostegids, Colbert continues as follows: "In the postcranial skeleton *Ichthyostega* showed a strange mixture of fish and amphibian characters. The vertebrae had changed but little beyond the crossopterygian condition, whereas in the caudal region *the fin rays of the fish tail were retained.* In contrast to the primitive vertebrae and the persistent fish tail, there were strong pectoral and pelvic girdles, with which were articulated completely developed limbs and feet, quite capable of carrying the animal on the ground" (Colbert 1980, 75). Colbert's picture is clear. *Ichthyostega* is a descendant of the crossopterygian fishes, modified in a way that made possible the invasion of the land. (For more detail, see Romer 1966, 87–90.)

Another important evolutionary transition is the emergence of mammals from a group of reptiles known as *Therapsids.* These reptiles were distinguished by an opening in the skull behind the eye and the development of a secondary palate separating the nasal passage from the mouth. Their skeletons also show some differentiation of types of vertebrae (as is found in mammals), and the limbs become "tucked in" under the body, rather than splayed out in the characteristic reptilian fashion. Because the fossil record is relatively rich, and there are a number of discernible trends, paleontologists debate the question of exactly where the reptile-mammal boundary should be drawn. However, one particular feature has been popular for diagnostic purposes: Mammals are commonly distinguished from reptiles by the criterion that, in mammals, the jaw joint is formed between the *dentary* and the *squamosal* bones; the reptilian jaw joint is formed between two different bones (the *quadrate* and the *articular*)—in mammals, these latter bones form part of the structure of the auditory apparatus.

Finally, let us look briefly at the reptile-bird transition. Although the fossil record of birds is very sparse (for reasons already noted), there are two well-preserved skeletons of early birds (and three other fragments). The skeletons come from the same deposit. Colbert expresses the gratitude of paleontologists for these findings: "This fine-grained rock [the Solnhofen limestone of Bavaria] . . . was evidently deposited in a shallow coral lagoon of a tropical sea, and fortunately for us flying vertebrates occasionally fell into the water and were buried by the fine limy mud, to be preserved with remarkable detail" (Colbert 1980, 182–183). These fossil specimens, of *Archaeopteryx,* are so similar to reptiles in skeletal features that, but for the unmistakable impressions of feathers in the rocks, they would have been identified as reptiles. The skull is a typical reptilian skull, a reptilian tail is present, and so forth. Colbert sums up the situation as follows: "Here was a truly intermediate form between the reptiles and the birds. The skeleton alone was essentially reptilian, but with some characters tending strongly towards the birds. The feathers, on the other hand, were typical bird feathers, and because of them *Archaeopteryx* is classified as a bird—the earliest and most primitive member of the class" (Colbert 1980, 183).

Gish has objections to all of these cases. One basic tactic is to ignore the similarities between ancestral and descendant forms and to concentrate on their differences. So, for example, Gish writes, "There is a tremendous gap, however, between the crossopterygians and the ichthyostegids, a gap that would have spanned many millions of years and during which innumerable transitional forms should reveal a slow gradual change of the pectoral and pelvic fins of the crossopterygian fish into the feet and legs of the amphibian, along with loss of other fins, and the accomplishment of other transformations required for adaptation to a terrestrial habitat" (Gish 1979, 78–79). Gish is quite correct to recognize that there is a difference in the limbs of the first amphibians and the fins of the most advanced crossopterygian fishes. Despite the structural similiarity between them, the amphibian limbs (and limb girdles) are larger and sturdier. On the other hand, the skull and vertebral differences between crossopterygian fishes and ichthyostegids are slight. Indeed, we can trace a progression of skeletal characteristics through an advanced group of crossopterygian fishes (the *rhipidistians*) toward the ichthyostegids. What Gish does is to demand a sequence of fossil forms that is continuous with respect to one set of characters *that he has chosen*. What nature supplies is a sequence that is continuous with respect to a different set of features. Given the vicissitudes of fossilization, there is no reason to expect a sequence of fossils showing continuous modifications of any characteristic we choose, even if that characteristic was continuously modified. Paleontologists think themselves lucky to be able to trace the continuous emergence of *some* characteristics.

Because the reptile-mammal transition is very richly documented, the tactic of focusing on differences is less easy to apply to it. However, Gish tries valiantly. He begins his discussion by noting that mammals are usually distinguished from reptiles by the formation of the jaw joint and the bony structure of the inner ear. Inspired by this, Gish constructs his own vision of what transitional forms there ought to have been: "There are no transitional forms showing, for instance, three or two jaw bones, or two ear bones. No one has explained yet, for that matter, how the transitional form would have managed to chew while his jaw was being unhinged and rearticulated, or how he would hear while dragging two of his jaw bones up into his ear" (Gish 1979, 85). Three points are made, and all are hopelessly wrong. First of all, evolutionary theorists are in no way committed to the idea that the two bones forming the reptilian jaw joint must have migrated *separately* so as to form part of the auditory structures. That is purely a figment of Gish's imagination. *He* speculates about the character of transitional forms, and then chides paleontologists because they do not find what he demands.

Second, there is a very clear explanation of how the transitional forms managed to chew (Crompton and Parker 1978; Kurtén 1971, 25–26; Crompton and Jenkins 1979; Colbert 1980, 136–137). The therapsids show a trend toward reduction of the bones that form the reptilian jaw joint (the quadrate and articular) and toward enlargement of the bones that form the mammalian jaw joint (the dentary and squamosal). The trend culminates in animals—one of which is aptly named *Diarthrognathus*, or "double jaw joint"—*in which both the reptilian and the mammalian jaw joints are present*. Gish's sympathy for therapsids and their feeding problems is thus misplaced. The transitional sequence we have provides good evidence that the reptilian jaw joint was not "unhinged" until the mammalian jaw joint had been formed.

Finally, there is no reason to worry about the hearing of advanced reptiles or early mammals. In reptiles, sound is frequently transmitted from the ground through bones in the jaw. Fossil evidence reveals that, in the early mammals, the tympanic membrane (the principal membrane of the ear) is quite low. This suggests that, as the quadrate and articular were released from their function in the reptilian jaw, refinement of the reptilian system of sound transmission was a gradual process. From the elements of the reptilian system for transmitting vibration, the mammalian auditory system was built. The early mammals show us an intermediate stage in the building of this system, a stage at which the quadrate and articular have been moved upward, but have not yet assumed their final position. Obviously, we do not have to assume that there was any stage of the process at which some previously functioning sensory system became dysfunctional. (For a much more refined discussion, see Crompton and Parker 1978, 195–198; Crompton and Jenkins 1979, 66.)

Hence the attempt to exploit differences between therapsids and mammals fails utterly. However, Gish has more tricks up his sleeve. The next try is to argue against "the idea that the 'mammal-like' reptiles were really mammal-like" (Gish 1979, 85). Gish's argument consists of several quotes from an article (Crompton and Parker 1978) that provides considerable insight into the evolution of the mammalian jaw. This would appear to be an unpromising source for Gish to use. And indeed it is.

The passages that Gish cites point out that certain features common to mammals are absent in advanced therapsids. The features in question concern the structure of the ear, the mode of tooth replacement, and the mechanics of the jaw. On this basis, Gish announces that therapsids were not mammallike. However, his conclusion is totally unwarranted. The standard way of separating therapsids from mammals is to use features like jaw mechanics and ear structure diagnostically. This means that no *therapsid* can be like a mammal in these respects: if it were like a mammal in these respects, then it would *be* a mammal. To claim that the therapsids are like mammals is to assert that, although they differ from mammals in not having the diagnostic characteristics of mammals, they show a trend toward *these and many other* mammalian features. The claim is amply confirmed by the fossil record.

An analogy may help us to understand Gish's trick. Imagine that we have a sequence of color patches, forming a continuous gradation from yellow through orange into red. For purposes of convenience, we divide them into two classes, the red ones and the yellow ones, by choosing some criterion of demarcation. (We might pick, for example, some wavelength in the middle of the orange range.) It would be absurd to object that we do not have a continuous sequence of colors, on the grounds that all the patches in the yellow class fail to satisfy the criterion for being in the red class.

The analogy is easily applied to the reptile-mammal transition. The fossil record supplies a sequence of organisms showing gradual changes in a number of characteristics. For taxonomic purposes, zoologists want to split these organisms into two classes, the reptiles and the mammals. Paleontologists are aware that they are imposing a division on a continuum:

Because, in the ictidosaurs, the transformation of the quadrate and articular bones had not taken place, these animals can be placed arbitrarily within the reptiles. All of which indicates how academic is the question of where the reptiles end and the mammals begin. (Colbert 1980, 137, also 246)

Rather there are one-quarter mammals, half-mammals, and so on. These transitional forms, representing several different populations becoming more mammal-like, are found in the late Triassic and the early Jurassic. (Kurtén 1971, 25)

The series is divided by using the criteria of jaw mechanics, ear structure, and features of the teeth. Gish argues that therapsids are unlike mammals because they do not have mammalian jaw mechanics and mammalian ear structure. Since it is exactly these criteria that are frequently used to separate otherwise similar animals, this line of attack is hopelessly naive.

However, Gish has one last salvo to fire at the reptile-mammal transition. He tries to show that the proposed transitional forms between major classes do not occur in the fossil record at the right times: "The known forms of *Seymouria* and *Didactes* [sic], which are said to stand on the dividing line between amphibians and reptiles, are from the early Permian. This is at least 20 million years too late, according to the evolutionary time scale, to be the ancestors of the reptiles" (Gish 1979, 87). Contrary to what one might expect from this account, evolutionary theorists do not make the absurd proposal that the transitional forms between amphibians and reptiles postdate the emergence of reptiles. Gish mistakes the thesis that *Seymouria* and *Diadectes* (not *Didactes*) are intermediate in structure between amphibians and reptiles, and sometimes taken to be relatively unmodified descendants of the organisms that linked amphibians and reptiles, with the claim that these animals are transitional forms. (In other words, a traditional evolutionary claim is that *Seymouria* shares a common ancestor with the reptiles, an ancestor that was a link in the evolutionary chain from amphibians to reptiles. Paleontologists used to believe that *Seymouria* is quite similar to this common ancestor, whereas the reptiles have diverged from the ancestral type.)

He then continues by making a further incorrect attribution: "According to evolutionists, mammals assumed supremacy over the reptiles at a relatively late time in reptilian history. If this is true, a reasonable assumption would be that the reptilian branch from which they arose developed late in the history of reptiles. Just the opposite is true, however, if the synapsids gave rise to mammals. The subclass Synapsida is dated among the earliest of known reptile groups, not the latest, and are [sic] supposed to have passed their peak even long before the appearance of dinosaurs" (Gish 1979, 87–88). Two points in this passage are correct. The mammals did assume supremacy comparatively late in reptilian history. Furthermore, the synapsids, from which the therapsids evolved, are reptiles that emerged relatively early. Yet no evolutionary theorist would agree with Gish's "reasonable assumption." The standard account, well documented by the fossil record, is that mammallike reptiles emerged fairly early. The advanced therapsids and the primitive mammals were apparently unable to compete successfully with the dinosaurs. During the heyday of the dinosaurs, the diversity of therapsids was greatly reduced, and the mammals that persisted were smaller in size than some of their ancestors had been. With the decline of the dinosaurs, the mammals began the explosive diversification that has produced the modern forms.

Gish alludes to this standard account in his parting shot at the reptile-mammal transition:

According to Romer, the synapsid reptiles dwindled in numbers during the Triassic, becoming essentially extinct by the close of that period, and many millions of years elapsed before their mammalian "descendants" rose to a position of dominance. If natural selection is one of the governing processes of evolution, and natural selection is defined as the process which enables the more highly adapted organism to produce the most offspring, then the above history, if true, seems to indicate that the reptile-to-mammal transition succeeded in spite of, rather than because of, natural selection. (Gish 1979, 88)

This paragraph adds further confusion. At the close of the Triassic, the diversity of the therapsids (the survivors of the synapsids) was greatly reduced. Some therapsid lines gave rise to mammals; others went extinct. More important, Gish's glancing reference to adaptation errs in neglecting that adaptation is *adaptation to an environment* and that the environment includes the competition. While the dinosaurs were ascendant, the early mammals were sufficiently adapted to their environment to survive, but not able to become dominant. (Recall the discussion of Wilder-Smith's nonweeding theorem. Evolutionary theory is not commmitteu to the idea that "advanced" organisms, or organisms that eventually become highly successful, must be able to defeat any competition in any environment.) After the dinosaurs became extinct, the mammals were able to occupy many vacant niches. Without competition from the dinosaurs, they diversified and became extremely successful.

Since I have dealt with all of Gish's arguments about the reptile-mammal transition, I shall discuss his treatment of *Archaeopteryx* only briefly. His tactic is one that we have seen before. Because *Archaeopteryx* had feathers, a diagnostic characteristic of birds, "It was not a halfway bird, it *was* a bird" (Gish 1979, 90). This is essentially the same move that Gish made in denying that the advanced therapsids were mammallike. Taxonomists draw a line to separate two classes of organisms. Gish then annnounces that there are no intermediate forms because everything falls on one side of the line or on the other. But this is a Pickwickian definition of *intermediate forms*. The question is whether the fossil record reveals organisms that have some charcteristics that are common in one class and other characteristics that are common in the other class. *Archaeopteryx* has a reptilian skeleton and the feathers distinctive of birds. The fact that, for convenience in taxonomy, biologists draw a line and include it with the birds does not abolish its kinship with the reptiles.

The examples discussed so far concern transitional sequences between classes of animals. None of them is as detailed as fossil links that connect smaller groups of animals. For example, there are very good fossil records for the evolution of rhinoceroses, horses, and tapirs from a common ancestor, for the evolution of crocodilians, and for the evolution of dogs and bears from a common ancestor. Gish recognizes that the existence of these sequences buttresses the evolutionary biologist's claim that, despite its incompleteness, the fossil record sometimes displays the evolution of two kinds of organism from a common ancestor. So he attempts to cast doubt on the most famous fossil sequence of all—the horse series.

At this point a new tactic emerges. Instead of offering a detailed argument, Gish simply appeals to authority. His sources are various. The maverick evolutionary theorist Richard Goldschmidt is quoted out of context. There is an appeal to a paper in the *Creation Research Quarterly*, which is said to reveal "many of the weaknesses and fallacies in the use of fossil horses as evidence for evolution" (Gish 1979, 101). But the true advance in argumentative technique lies in Gish's exploitation of the fact that paleontologists disagree about details. The fossil record of horses is very rich. There are many different forms, clearly related to one another, and the task of charting these relationships is complicated. Paleontologists do not question the relationships, and the overall line of evolution is uncontroversial. But, as in many other sciences, when the general form of the answer to a question is known, there is frequently considerable disagreement about the specifics. So, for example, Gish is able to quote David Raup: "By this I mean that some of the classic cases of Darwinian change in the fossil record, such as the evolution of the horse in North America, have had to be discarded or modified as the result of more detailed information—what appeared to be a nice simple progression when relatively few data were available, now appears to be much more complex and much less gradualistic" (Raup 1979, quoted in Gish 1979, 103). Torn out of context, Raup's remark may make it appear that paleontologists have given up the idea that the organisms in the horse sequence are related to one another. However, that is not the issue. No paleontologist doubts that there is a process of "descent with modification" that embraces all the animals preserved in the horse sequence. What is at issue is *how* they are related.

This tactic is obviously very powerful. Given any field of science, it is always possible to find experts who disagree about some technical question. Take their remarks out of context, and you can use them to question a statement on which all parties to the debate agree. As we shall see in the next chapter, Gish and other Creationists eagerly exploit current disputes about mechanisms of evolution. The same tactic also looms large in Gish's attack on the fossil record of human evolution (Gish 1979, chapter vi). The fact that paleontologists disagree on *how* fossil hominids are related to one another and to fossil apes is used to suggest that it is reasonable to deny *that* fossil hominids are related to one another and to fossil apes. Add a gleeful account of the infamous Piltdown hoax, and Gish can sow much confusion.

Evolutionary theorists believe that the fossil record, while imperfect, shows some examples of large-scale evolution. Gish's book discloses a number of ways to attack this idea: (1) Overlook the fact that there are independent reasons for believing both that the fossil record is incomplete and that it is biased. (2) Construct demands about what characteristics are to be gradually modified and then complain that the fossil record is incomplete in these respects. (3) Advance implausible speculations about what transitional forms would have been like, and then point out that they do not occur. (4) Ignore explanations given by evolutionary theorists about how the transition occurred. (5) Assert that intermediate forms belong either to the earlier taxonomic class or to the later taxonomic class, so that they cannot be truly transitional. (6) Find scientists who disagree about details, and quote them judiciously so as to portray them as questioning a fundamental point. Anybody who thinks that these tactics are legitimate will end up believing that the fossils say "No!" But nobody should favor such tactics, and nobody ought to accept Gish's conclusion.

In concluding this section, it may help to understand why I have quoted so freely from Gish if we look at some of the remarks of his fellow Creationists:

Birds, in the evolution model, came from reptiles. We must, therefore, accept that one day a bird was hatched from a reptile egg having, instead of reptilian scales, a complete set of perfectly developed feathers. Such a flair for imaginative tales, though very useful to the Brothers Grimm, cannot be tolerated in science. Our now superabundant fossil inventory does not contain even one fossil covered partially with feathers and partially with scales. (Hiebert 1979, 48)

No one has produced yet a single fossil with half-way wings or a fossil of an animal showing a transition between the cold-blooded scaled reptile and the warm-blooded feathered bird. . . . And not even the fossil *Archaeopteryx* can qualify as a transitional form, because it apparently had a bird-like skull, perching feet, and fully developed wings with feathers. (Moore 1974, 17)

Never are fossils of creatures found with incipient eyes, with half-way wings, with half-scales turning into feathers, with partially-evolved forelimbs, or with any other nascent or transitional characters. Yet there must have been innumerable individuals which possessed such features, if the neo-Darwinian model of evolutionary history is correct. (Morris 1974b, 91–92)

Apart from Hiebert's idea that the evolutionary process is supposed to go by leaps and bounds *and* to be gradual, and aside from Moore's factual errors about *Archaeopteryx*, what we have here is a broad application of tactics (1) and (3). *Creationists* imaginatively describe transitional forms, and then ridicule their own creations. It is a relief to turn from passages like these to Gish's relatively more competent discussion.

Worries about wings

In 1871 St. George Mivart published *The Genesis of Species*. This book was intended to expose the errors of the *Origin*. Like other early critics of Darwin, Mivart had a specific worry. How, he asked, are we to account for the development of complex structures, like the wings of birds and bats, or the eyes of numerous animals? No doubt, once formed, these structures are very useful to their possessors. But during the early stages of their supposed evolution, what advantages could they possibly have served?

Mivart's questions are echoed by contemporary Creationists, who dutifully illustrate them with the same examples:

But even if variation, or recombination, really could produce something truly novel, for natural selection to act on, this novelty would almost certainly be quickly eliminated. A new structural or organic feature which would confer a real advantage in the struggle for existence — say a wing, for a previously earth-bound animal, or an eye, for a hitherto sightless animal — would be useless or even harmful until fully developed. There would be no reason at all for natural selection to favor an incipient wing or incipient eye or any other incipient feature. (Morris 1974a, 53)

There are at least three serious objections to an evolutionary scheme here: a) How did creatures with such highly specialized structures and functions as we see everywhere in nature, and which are absolutely essential to their survival, manage to exist while these functions and structures were evolving? b) Why didn't natural selection eliminate such useless, imperfectly developed hindrances? c) Where are the nascent structures that creatures should be developing today for future use?

These problems are always given a wide berth in evolutionary literature. (Hiebert 1979, 106)

Hiebert is wrong. Evolutionary biologists provide extensive discussion of the origin of "evolutionary novelties." Their conclusions are summarized in a lucid article by Ernst Mayr (Mayr 1959; reprinted in Mayr 1976, 88–113).

As Mayr points out, there are three general ways of understanding the emergence of evolutionary novelties: "The new structure originates (1) as a pleiotropic by-product of a change of genotype, (2) as the result of an intensification of function, (3) as the result of a change of function" (Mayr 1976, 95). In other words, a new structure may arise because genotypes that are selected for their ability to produce quite different features happen to generate it. Or it may be produced because any increase in a particular ability is advantageous to an organism. Or the incipient structure may yield a different advantage from the final version. Whenever we can apply *any* of these explanations, there is no ground for supposing, as Morris does, that incipient structures are not advantageous or for asking Hiebert's question (b). Evolutionists are challenged to find plausible scenarios for the convergence of certain complex structures. Mayr's categories help them to respond. The secretion of nectar by plants may be an instance of (1). Thus it is possible that sweet liquid was originally secreted as a by-product of a genotype selected for quite different effects. A spectacular example of (2) is the modification of the feet of the ungulates (such as deer and antelopes) in response to selection for speed in running. Here, the fossil record even permits us to document the changes (if not their causes). In other cases, such as the famous example of the eye, a similar account appears plausible. We can explain the value of an incipient eye by supposing that there was an initial neural structure that could be affected by light and that selection favored organisms with ever greater sensitivity. In this case, biologists cannot document the process by appealing to the fossil record. But that is not the issue. The Creationists' challenge is meant to show that evolutionary theory is *incapable* of providing any account of the advantages accruing to incipient structures, for example, a rudimentary eye. That challenge is met by explaining how to construct a Darwinian history that would allow for the gradual emergence of the eye.

The most common explanation of evolutionary novelties appeals to *change* of function. An organ or structure originally develops because it enables those organisms that have it to secure a certain advantage. Once developed beyond a certain point, it can be put to what we may call "unanticipated uses." So it is frequently suggested that the structures that ultimately enabled birds to fly—wings, supporting bones, and musculature—initially emerged in the reptiles because their presence bestowed quite a different type of advantage. Feathers, for example, are not only useful for flying organisms. They can also play an important role in the control of body temperature. So feathers may originally have emerged in response to pressure for better thermoregulation. Feathered forelimbs, developed by reptilian ancestors of birds, might have played a useful part in a thoroughly earthbound existence. Thus, on one theory of the origins of flight, the scenario goes as follows: "The first birds were fast runners that flapped their feathered forelimbs to help them along, as do many modern birds that run rapidly, or to make leaps to catch insects. Gradually the wings enlarged through the processes of mutation and selection, so that in the end they became organs of flight rather than accessories to running" (Colbert 1980, 184–185). I want to emphasize that this is *one possible* scenario. To claim that it is the true Darwinian history of the development of the modern bird's wing would be premature and foolish. But no such claim is needed. Mivart and his Creationist successors maintain that, in cases like this, no Darwinian history is *possible*. Hence we are only required to explain how such a case could be handled given the resources of evolutionary theory.

The answer to Morris's point and to the analogous point raised by Hiebert is thus straightforward. Complex structures can be formed gradually through natural selection if, from the beginning, they can meet any of three conditions: (1) They are linked to an advantageous character; or (2) They secure some degree of the advantage they will ultimately confer; or (3) They bring a quite different advantage. Different explanations apply to different cases. Some explanation can be given for each of the allegedly troublesome cases.

Finally, Hiebert's subsidiary questions do not raise any serious problem. The third question asks evolutionary theory to predict the future evolution of contemporary organisms. We have already seen why such prediction is impossible. His first question presupposes a false view of the evolutionary process. How did bats continue to exist before they had wings? How did they manage while the wings were developing? The answer is very obvious. The ancestors of the bats occupied a different place in the environment. Once the bats had evolved they were able to exploit environmental space that was inaccessible to their ancestors. And they have been very successful ever since. Hiebert's question makes the naive assumption that the ancestors of current organisms, ancestors who lack certain equipment necessary for their descendants to use particular environmental resources, would have occupied exactly the same place in the environment as their fortunate descendants.

Goodbye Paluxy

All the Creationists' major objections have now been examined. What appear, at first glance, to be imposing obstacles turn out, on closer inspection, to be conjuring tricks employing inaccuracy, misrepresentation, dazzling numbers, and layers of confusion. As a teacher, I sometimes wonder after an exam why some students became confused on a particular point, and I try to understand how my presentation or the formulation in a book could have misled them. In the case of the Creationist authors we have studied, there is no great difficulty in seeing how muddles arise. They want to use scientific data and scientific principles to attack evolutionary theory. So they skim, searching for ammunition. When they find a claim that seems to be at variance with evolution, they seize it as a trophy to bring back to the Institute for Creation Research for public display. If they actually tried to understand the terrain they scavenge, they would have learned some interesting science. Instead, they seem to acquire only the most tenuous grasp of complex theories and then offer their muddled caricatures of important scientific works to as wide an (inexpert) audience as they can reach. (It is possible, of course, that their understanding is greater than that revealed in their confused discussions. But I am loath to accuse them of perverting ideas that they actually comprehend.)

Having surveyed the "mountains," we may turn to the molehills. Let me begin with a general point. Every scientific theory faces problems at all times. There are always details to work out, interesting cases to which the problem-solving strategies of the theory have not yet been applied, puzzling examples where straightforward application of those strategies seems to give the wrong answers. The theory is articulated and refined as scientists try to tackle the outstanding problems. Because of its record of successful problem solution, the theory holds out the promise that, with sufficient ingenuity, outstanding difficulties can be resolved.

So it is with evolutionary theory. There are numerous specific questions, questions of evolutionary relationships and geographical distribution, for example, to which we lack detailed answers. If the Creationists were to point out that among these questions are some that have defied the resources of the best evolutionary theorists over a significant period of time, then we might begin to wonder whether something is amiss with evolutionary theory. Ironically, when Creationists make specific objections, they do not focus on problems that have been tried and found recalcitrant. Instead, they report findings of their own that they claim cannot be integrated within the framework of evolutionary biology.

Evolutionary theorists do not become terribly excited by these reports. They do not drop their current research and hasten off to Rhodesia, Arizona, or wherever the latest devastating phenomenon has been unearthed. Why not? Is their attitude simply one of arrogance? Or laziness? Or cowardice? Creationists would like the world to believe that it is. They have crucial counterevidence, they claim, to which the elitist scientific establishment will not listen: "Evolutionists begin with the assumption that evolution is true, and from there they gather data and assemble charts to fit this preconception. Facts counter to the geologic column are considered in error, explained through geologic mechanisms like faulting or thrusting, whether or not there is supporting physical evidence, or simply shelved as quirks of nature" (Wysong 1976, 392).

To understand the attitude of evolutionary theorists we need to appreciate the kinds of problems they are being asked to solve. The issues in question are of absolutely no theoretical interest. The Creationist tells us that there are cave pictures in Rhodesia that depict dinosaurs (Morris 1972, 31) or shows us a photograph of such pictures (Wysong 1976, 380). There is an obvious explanation. The picture is a crude representation of some other animal. Next the Creationist reports on places where geologically "older" sediments are found above "younger" deposits. Geologists greet this news with a yawn. They know very well that "overthrusting" is possible, and contemporary geology (specifically the study of tectonic movements) gives a general explanation of the phenomenon. (Strata can be deformed and older rocks thrust over younger rocks. Geologists have independent ways of testing when this has occurred.) Nobody is interested in working out another application of the theory. There are too many genuinely important problems to be tackled.

So we come to Paluxy, and the Creationists' favorite example. The Paluxy river bed, in central Texas, is alleged to contain "large numbers of both dinosaur and human footprints" (Morris 1974a, 122; see also Wilder-Smith 1981, 96–98; Wysong 1976, 373–377). This would shake the foundations of evolutionary theory, because, of course, the dinosaurs are supposed to have been long extinct by the time the hominids arrived on the scene. Evolutionary theorists reply that the alleged human footprints are not genuine. In fact, it is sometimes charged that the tracks have been modified. These charges provoke Wilder-Smith to some accusations of his own: "G. G. Simpson labels them [the supposed tracks] as a plain lie. The authenticities of the discoveries at Glen Rose [Paluxy] would in one blow absolutely and radically destroy Simpson's lifetime work as a proponent of Neodarwinism. The well known publishing scandals involving Macmillan and the suppression of Velikowski's research, as well as the Piltdown hoax, provide us with much food for thought in respect to the publication of facts which would correct erroneous generally accepted scientific philosophies" (Wilder-Smith 1981, 98).

There is utterly no reason to believe that Simpson is a villain, desperately conniving to prevent the unwelcome truth from emerging. He is a sincere and dedicated scientist, who has shown himself able to retract and modify his views in response to informed and responsible criticism. Moreover, he like other scientists, has heard the Creationists cry "Wolf!" too often.

However, because Creationists have placed so much emphasis on the Paluxy findings, a number of trained scientists have gone to take a look. Their first-hand observations have exposed some interesting features of the Paluxy situation. Many of the tracks are not readily distinguished, even by expert paleontologists. Yet more interesting is the disclosure of a piece of social history. During the Depression, a few local inhabitants made money by carving tracks in pieces of rock. Because of these unwelcome discoveries, some Seventh-Day Adventists have abandoned the Paluxy findings as evidence for Creationism. Even John D. Morris, the son of Henry Morris, and himself a member of the Institute for Creation Research, has been forced to concede that some of the tracks are counterfeit. Nevertheless, "scientific" Creationists continue to insist that Paluxy yields serious evidence against evolution. The most recent printings of Creationist books do not intimate that much of the "Paluxy data" is already recognized as dubious by Creationists themselves—and that even people who share their theological views reject it wholesale.

The rewards of going forth to confirm orthodox explanations of Creationist "findings" are slight. If scientists do not investigate, then they are assailed with complaints that they do not attend to the counterevidence that would topple evolutionary theory. When they do answer the call, they are accused of making biased observations and failing to see what is really there. Even if the tracks at Paluxy finally disappear from the Creationist literature, there will always be new Paluxies. Creationists will raise pedestrian problems of no theoretical interest, and they will "discover" new peculiarities in the fossil record. The fact that they are so obviously grinding their pet axes generates concern about the purity of their findings. To go out and verify that standard applications of evolutionary theory will take care of the problematic phenomena takes time from more profitable research. To go out and show that some of the alleged findings are counterfeit only leads Creationists to proclaim more loudly that the rest are genuine and that prejudiced scientists look only at what supports their preconceptions. The facts that Creationists cite are not *inexplicable* by evolutionary theory. They are simply *unexplained*. Ironically, they remain unexplained because they are not sufficiently problematic to merit further attention. So, the invitation to chase wild geese is usually declined.

5

What Price Creationism?

Three types of Creationism

Equal time for equally good theories sounds like a good maxim for educators. We have subjected evolutionary theory to intense examination. All kinds of objections have been considered, but the theory has emerged unscathed. We have seen, among other things, how evolutionary theory has led to the development of new and successful scientific disciplines and how it has been able to explain, in some detail, phenomena as specific as the peculiarity of the mammals of Madagascar (to cite one of the countless concrete examples of problem-solving success). Now it is time to look at Creation "science." As we shall quickly discover, the first problem is to find any positive theory at all. We shall then subject what there is of a theory to *some* of the questions we have asked of evolutionary theory. Does it offer us a set of problem-solving strategies that enable us to unify diverse biological observations? Does it make itself vulnerable to possible refutation? It will turn out that we cannot ask this theory many other questions. There is no point in asking, for example, what outstanding research problems Creation "science" has or what special experimental techniques it introduces. Nor can we inquire about the next likely area of breakthrough or request information about which parts of the theory are better confirmed than others. We cannot raise these questions for "scientific" Creationism because the "theory" that Creationists offer is not sufficiently like genuine science to make sense of such routine inquiries. This chapter will draw a simple moral: People who live in Creationist houses should not throw methodological stones.

Before turning to "scientific" Creationism itself—that is, the theory peddled by the Institute for Creation Research and advertised by the Moral Majority—it is important to distinguish two other forms that

a belief in special creation might take. The central idea of strict Creationism is that all kinds of organisms presently existing, and perhaps
some more, were formed on the earth in a single event. Some people
hold this view purely as an article of religious faith, making no claim
that it is a part of science supported by scientific evidence. Such people
accept strict Creationism because its central doctrine follows from two
other beliefs that they hold: (i) The Bible is to be read literally; (ii) When
the Bible is read literally, it says that all kinds of organisms were
formed on the earth in a single event. I disagree with this view, because
I do not believe it is possible, let alone reasonable, to read the Bible
literally. (Nor do I think that Christians and Jews are compelled, as
sincere believers, to read the Bible literally; see chapter 7.) However,
I have no intention of criticizing Creationism insofar as it is held as
an explicitly religious belief, a belief that is recognized as running
counter to the scientific evidence.

There is another way to be a Creationist. One might offer Creationism
as a scientific theory: Life did not evolve over millions of years; rather
all forms were created at one time by a particular Creator. Although
pure versions of Creationism were no longer in vogue among scientists
by the end of the eighteenth century, they had flourished earlier (in
the writings of Thomas Burnet, William Whiston, and others). Moreover,
variants of Creationism were supported by a number of eminent nineteenth-century scientists—William Buckland, Adam Sedgwick, and
Louis Agassiz, for example. These Creationists trusted that their theories
would accord with the Bible, interpreted in what they saw as a correct
way. However, that fact does not affect the scientific status of those
theories. Even postulating an unobserved Creator need be no more
unscientific than postulating unobservable particles. What matters is
the character of the proposals and the ways in which they are articulated
and defended. The great scientific Creationists of the eighteenth and
nineteenth centuries offered problem-solving strategies for many of
the questions addressed by evolutionary theory. They struggled hard
to explain the observed distribution of fossils. Sedgwick, Buckland,
and others practiced genuine science. They stuck their necks out and
volunteered information about the catastrophes that they invoked to
explain biological and geological findings. Because their theories offered
definite proposals, those theories were refutable. Indeed, the theories
actually achieved refutation. In 1831, in his presidential address to
the Geological Society, Adam Sedgwick publicly announced that his
own variant of Creationism had been refuted:

Having been myself a believer, and, to the best of my power, a
propagator of what I now regard as a philosophic heresy . . . I think

it right, as one of my last acts before I quit this Chair, thus publicly to read my recantation.

We ought, indeed, to have paused before we first adopted the diluvian theory, and referred all our old superficial gravel to the action of the Mosaic Flood. For of man, and the works of his hands, we have not yet found a single trace among the remnants of a former world entombed in these ancient deposits. In classing together distant unknown formations under one name; in giving them a simultaneous origin, and in determining their date, not by the organic remains we had discovered, but by those we expected hypothetically hereafter to discover, in them; we have given one more example of the passion with which the mind fastens upon general conclusions, and of the readiness with which it leaves the consideration of unconnected truths. (Sedgwick, 1831, 313–314; all but the last sentence quoted in Gillispie 1951, 142–143)

Since they want Creationism taught in public schools, contemporary Creationists cannot present their view as based on religious faith. On the other hand, the doctrine is too dear to be subjected to the possibility of outright defeat. What is wanted, then, is a version of Creationism that is not vulnerable to refutation, but that appears to enjoy the objective status that can only be conferred by evidential support. This is an impossible demand. A theory cannot drink at the well of evidential support without running the risk of being poisoned by future data. What emerges from the conflict of goals is the pseudoscience promulgated by the Institute for Creation Research. It is vaguely suggested that the central Creationist idea could be used to solve some problems. But the details are never given, the links to nature never forged. Oddly, "scientific" Creationism fails to be a science not because of what it says (or, in its "public school" editions, very carefully omits) about a Divine Creator, but because of what it does not say about the natural world. The theory has no infrastructure, no ways of articulating its vague central idea, so that specific features of living forms can receive detailed explanations.

Despite my best efforts, I have found only two problem-solving strategies in the writing of "scientific" Creationists. Most of the literature is negative, a set of variations on the antievolutionary themes we have heard in the last three chapters. The positive proposals of Creation "science" are remarkably skimpy. Even *Scientific Creationism* (the work that is intended to enable teachers to present the "creation model") spends far more pages attacking evolutionary theory than in developing the Creationist account. Nevertheless, there are passages where a positive doctrine seems to flicker among the criticisms. Similarly, the much earlier book *The Genesis Flood* (Whitcomb and Morris 1961), mixes attempts at constructing Creation "science" with its explicit Biblical interpretation. Because of the uniformly negative character of most other Creationist writings, my evaluation of the positive theory presented by "scientific" Creationists will be based primarily on these two works. The two problem-solving strategies of "scientific" Creationism are the attempt to use Flood Geology to answer questions about the ordering of fossils and an appeal to a mix of "design" and historical narrative to account for the properties, relationships, and distributions of organisms. I shall document my remarks about pseudoscience by taking a closer look at how these explanatory vehicles operate.

Room at the top for the upwardly mobile?

Creationists recognize that the fossil record is ordered. All over the earth we find a regular succession of organisms through the rock strata. At the lowest levels, we find only small marine invertebrates. As we move up, other groups of animals are encountered: Fish join the marine invertebrates; then come the amphibians, reptiles, and, finally, the mammals and birds. Of course, within each of these groups there is also an order. The first reptiles diversify, giving rise at later times to several different groups, some of which, like the dinosaurs, die out, while others of which, like the snakes, turtles, lizards, and crocodilians, persist to the present. Even without evolutionary assumptions, it is possible to offer a simple explanation of this order. The different strata all contain remains of the distinctive organisms that were alive at the time at which those strata were deposited, and the order reflects the fact that different groups of animals have existed at different times. In other words, the animals who have inhabited the earth were not all contemporaries.

Although this simple explanation makes no commitment to the *evolution* of organisms, Creationists cannot accept it. For they believe that all the animals that have ever existed were formed in one original event of Creation. Nor can they abandon this belief without forswearing the theological payoff of their "science." So how are they to explain the order of the fossil record? Some antievolutionists of the late nineteenth century ascended to new levels of ad hoc explanation with two transparent ruses: (i) The Devil placed the fossils in the rocks to deceive us; (ii) God put the fossils there to test us. Contemporary Creationists are more subtle. They invoke the Flood. They hypothesize a worldwide, cataclysmic deluge, which destroyed virtually all the animals of the earth and deposited almost all the fossil-bearing rocks, thereby producing the fossil record.

Here is an outline of their major ideas. The primeval earth was a very pleasant place, consisting of land masses divided by "narrow seas." It was surrounded by a canopy of water vapor, producing a "greenhouse effect." In the Flood, water came from two directions; primitive waters inside the earth burst through the crust, and, at the same time, the vapor canopy was broken up to cause torrential rain. Some humans and animals escaped in boats (including, presumably, Noah, his family, and a pair of [land] animals of every kind). Others, less lucky, were drowned or destroyed, and some of them were engulfed by mud and other debris that were later deposited as sediments. (The latter animals are the ones that became fossilized.) Finally, the Flood came to an end, partly as the result of evaporation, partly because mountains erupted, producing basins in which the residual waters were entrapped. At this point, the remaining animals dispersed, bred, multiplied, and spread themselves over the new earth.

The attempt to explain geological formations by reference to the Flood is not new. Contemporary Creationists are heirs to a long tradition, begun by Thomas Burnet (whose *Sacred Theory of the Earth* was published in 1681). The idea of invoking a *single* cataclysm (and a single period of Creation) had been abandoned by the wiser geologists by the end of the eighteenth century, at which time the enterprise appeared hopeless. Nineteenth-century Catastrophists—Cuvier, Buckland, Sedgwick, and their followers—preferred to think of Noah's Flood as the last in a series of catastrophes. But "scientific" Creationists will have none of these newfangled compromises. So their account is vexed by all the questions that arose for their illustrious predecessors— as well as other questions inspired by the discoveries of the past 150 years. Here are just a few. How exactly did the land reemerge? (In traditional terms, how was "the pond drained"?) Did *all* kinds of land animals go on Noah's ship? If so, why are there so many kinds that are unrecorded in "post-Deluge" strata? How were the domestic economies managed during the voyage of the Ark? Obviously, the *mechanisms* of the whole episode could stand considerably more description. Creationists sometimes admit the point; Morris exhorts teachers to prepare geology students who can help with the task of working out the details (Morris 1974a. 129–130). However, neither he nor any other Creationist I have read seems to have any definite conception of where or how to begin this Herculean task.

Those disposed to take a charitable attitude toward Flood Geology might offer the following appraisal. Flood Geology is really only the sketch of a theory. Much further work is required to make a full-fledged theory. Still, in 1859, Darwin presented only a sketch of a theory, so really the sketchiness of Flood Geology does not doom the Creationist enterprise. They simply need time.

This apparently apt analogy comes apart when one compares Darwin's offering with the wares of the Institute for Creation Research. Darwin's theory and Flood Geology (in its contemporary version) are both "theory sketches" only in the same sense that a Leonardo drawing and a Rorschach inkblot are both sketches of works of art. Unlike Darwin's theory, Flood Geology offers no detailed explanation of any aspect of the fossil record. It provides nothing even remotely analogous to Darwin's studies of barnacles and South American mammals. Unlike Darwin's theory, Flood Geology shows no promise of fruitful interchange with other sciences. Darwin built upon the existing science of his day, and his proposals stimulated fruitful work throughout biology and geology. By contrast, as we shall see below, "scientific" Creationism would force us to abandon well-developed related theories, without any hint of how to construct replacements. Another important difference is that Flood Geology, as currently practiced, does not aim at advancing science — it does not seek to extend the range of phenomena that are open to scientific investigation. The goal is to fight a rearguard action, not to open up new areas of inquiry. Moreover, as we shall see, Flood Geology is very vague. Where Darwin explained carefully how he proposed to address biological questions, making clear how the application of his problem-solving strategy was to be made, Flood Geology is indefinite. Finally, Flood Geology is not a new proposal, daring and untried. There is little room to think that further efforts will enable the "theory" to emerge as a detailed solution to the problems it hopes to resolve. If Creationists do succeed in luring some talented geologists to their cause, I am inclined to mourn the waste of ability. Their program has no more to offer than it had when Thomas Burnet wrote the original script.

To see why Flood Geology deserves an obituary, let us watch it in action. The most ambitious attempt at detailed problem solving is presented by Henry Morris. Morris is very emphatic that Flood Geology accounts for the order of the fossils. After announcing fourteen "obvious predictions" of his story, he concludes, "Now there is no question that all of the above predictions from the cataclysmic model are explicitly confirmed in the geologic column. The general order from simple to complex in the fossil record in the geologic column, considered by evolutionists to be the main proof of evolution, is thus likewise predicted by the rival model, only with more precision and detail. But it is the exceptions that are inimical to the evolution model" (Morris 1974a, 120). Bold words. Before we take up the large claims made for Flood Geology, let us consider the swipes at the evolutionary account. First, the "general order from simple to complex" in the fossil record is not considered "the main proof" of evolution. Evolutionary theory rests on its ability to subsume a vast number of diverse phenomena — including the *details* of biogeography, adaptive characteristics, relationships among organisms, and the sequence of fossils — under a single type of historical reasoning. To say that "the general order from simple to complex" in the fossil record is the primary evidence for evolution is like saying that the fact that most bodies tend to fall is the primary evidence for Newtonian theory. Second, those "inimical exceptions" are our old friends the overthrusts, the cave drawings — and, of course, Paluxy. As I have already pointed out, these are not genuine problems for evolutionary theory.

How exactly does Morris propose to "predict" the order of the fossils from his model? Let us look at some of his "predictions" and their justifications:

3. In general, animals living at the lowest elevations would tend to be buried at the lowest elevations, and so on, with elevations in the strata thus representing elevations of habitat or ecological zones.

4. Marine invertebrates would normally be found in the bottom rocks of any local geologic column, since they live on the sea bottom.

5. Marine vertebrates (fishes) would be found in higher rocks than the bottom-dwelling invertebrates. They live at higher elevations and also could escape burial longer.

6. Amphibians and reptiles would tend to be found at still higher elevations, in the commingled sediments at the interface between land and water. . . .

9. In the marine strata where invertebrates were fossilized, these would tend locally to be sorted hydrodynamically into assemblages of similar size and shape. Furthermore, as the turbulently upwelling waters and sediments settled back down, the simpler animals, more nearly spherical or streamlined in shape, would tend to settle out first because of lower hydraulic drag. Thus each kind of marine invertebrate would tend to appear in its simplest form at the lowest elevation, and so on.

10. Mammals and birds would be found in general at higher elevations than reptiles and amphibians, both because of their habitat and because of their greater mobility. However, few birds would be found at all, only occasional exhausted birds being trapped and buried in sediments.
(Morris 1974a, 118–119)

It is hard to know where to start. Morris appears to have three possible explanatory factors: (1) *habitat* (lower dwelling animals were deposited first), (2) *hydraulic characteristics* (the order of deposition depends on the animal's resistance to the downward waters), (3) *mobility* (more mobile animals will be deposited later). The passages I have quoted juggle these three methods so as to obtain the desired results.

Now, for all the extravagant claims about "prediction" with "more precision and detail," the account Morris offers is extremely vague. Puzzles begin to appear in large numbers when we start to consider the details of the fossil record. Why are bottom-dwelling marine invertebrates found at *all* levels of the strata? Why are some very delicate marine invertebrates, which would have been likely to sink more slowly, found at the very lowest levels? Why are all the "modern" fishes (the *teleostean* fishes, which, on the standard account, emerged only in the age of dinosaurs and became spectacularly successful as the reign of the dinosaurs came to its end) found only in Morris's "late Flood" deposits? Why do these particular fishes not occur, as other fishes do, at lower levels? Why are whales and dolphins only found at high levels, while marine reptiles of similar size are found only much lower? Why do lumbering creatures like ground sloths appear only in Morris's post-Flood deposits, while much more agile mammals (such as the ancestors of contemporary carnivores and ungulates) appear much lower? Why are the *flying* reptiles found "in the commingled sediments at the interface between land and water"? Why were not *most* of the birds "exhausted," since perching places would have been hard to find in the raging Deluge? The sequence of questions could go on and on.

Morris does not consider the particularities. So the idea that we get *more* detail from his account is simply bluster. In fact, given that the problems raised for evolutionary theory are spurious—the "inimical exceptions" do not present any difficulty—the question we must ask is whether Flood Geology can *emulate* the ability of evolutionary theory to explain the fossil record. There are two ways in which Creationists might elaborate their proposals. One would be to acknowledge that the account so far given is programmatic and incomplete and to face up to the task of working out the details. The second would be to deny that there are residual questions for "scientific" Creationism to answer.

In the spirit of Morris's exhortation to young paleontologists, Creationists might concede that Flood Geology is only a "sketch." However, it is hardly a matter of adding a bit of detail to the main lines of the account. Problems are everywhere, solutions nowhere. What were the mechanics of the Flood? How were the animals preserved? Why are the details of the fossil ordering as we find them to be? It is reasonable to wonder what Flood Geology does have going for it that inspires people to work further on it. Wonderment increases when we realize that Creationists have abandoned the position of the most enlightened nineteenth-century Catastrophists. (For a lucid account of early-nineteenth-century Flood Geology, see Gillispie 1951.)

Here is a different "theory sketch" about the history of life on earth. For a very long time, the earth has been a laboratory for clever aliens who live in outer space. Periodically, they have "seeded" our planet with living organisms. In the beginning, they were only able to produce rather simple terrestrial organisms. So they started off with some marine invertebrates. After a while, they came for a visit to see how things were going. At this time, a dreadful thing happened. Something about the alien spacecraft caused a cataclysm on the earth; volcanoes erupted, there were massive earthquakes, enormous tidal waves, and so forth. Perhaps the spacecraft emitted some peculiar type of radiation that triggered all these unfortunate events. In any case, all the first crop of organisms was destroyed and buried in the cataclysm. The whole experiment was spoiled. However, since they understood the moral of Kipling's "If," the aliens decided to try again. Their technology had now improved, and they were able to manufacture more complicated animals. Some of those that had not performed well on the first round were dropped from the cast. The experiment went very well again—until they came to take another look. Once again, their presence led to disaster, and they were forced to start over from scratch. So it has gone for a number of trials. (How many would you like?) The aliens are very persevering. They still have not figured out what it is about their presence that makes the earth go into convulsions. But their technology is clearly improving by leaps and bounds. After all, last time they made us.

My brief acquaintance with the "theory sketch" of the last paragraph has not yet led to a deep attachment. (It took me about ten minutes to make it up.) So I shall not exhort others to join me in working out the details. However, I do want to suggest that, *from a scientific point of view*, my silly proposal is no worse than Morris's Flood Geology. It would not be difficult to mimic the "fourteen predictions." ("What evolutionists call *trends* are really the aliens' progress in fine tuning already workable designs." Notice, too, that my theory, like Morris's vague account of Flood Geology, has plenty of "wiggle room.") The point of the comparison should be obvious. There is utterly no reason to take either my proposal or Flood Geology seriously—or to exhort promising students to waste their careers in the pursuit of such obvious folly.

The second way in which the Creationists can respond to questions about the details is even worse. Instead of taking the problems of detail seriously, they can contend that we can never know how the Flood worked. All that can be done is to lay down some general considerations, which hold "as a rule," and suggest that, given some unknowable distribution of upwelling waters, torrential rains, and trapped animals, everything sorted out as it did. So, for example, the questions I have raised about teleostean fishes, whales, flying reptiles, and giant sloths can just be ducked. These are "the exceptional cases."

Some passages suggest that, when push comes to shove, Morris and his fellow Creationists will slide in this direction:

14. All the above predictions would be expected statistically but, because of the cataclysmic nature of the phenomena, would also admit of many exceptions in every case. In other words, the cataclysmic model predicts the general order and character of the deposits but also allows for occasional exceptions. (Morris 1974a, 120)

In other localities, and perhaps somewhat later in the period of the rising waters of the Flood, in general, land animals and plants would be expected to be caught in the sediments and buried; and this, of course, is exactly what the strata show. Of course, this would be only a general rule and there would be many exceptions, as currents would be intermingling from all directions, particularly as the lands became increasingly submerged and more and more amphibians, reptiles and mammals were overtaken by the waters. (Whitcomb and Morris 1961, 275)

The remarkable point about these passages is not the *number* of qualifying phrases, but their *variable strength*. Are there "many exceptions in every case" or are the exceptions "occasional"? I do not know what Morris or Whitcomb intends. Yet one thing is clear. Such passages can be used to maintain that the anomalies I have mentioned are not genuine problems for Flood Geology. Morris and Whitcomb have carefully provided an all-purpose escape clause. So while (alleged) exceptions are "inimical" to evolutionary theory—they would mean refutation—exceptions, even hordes of exceptions, in no way weaken the case for Creation"science." For Creation "scientists" data has only one function; it is a potential source of problems for evolution. Counterexamples to the "theory" of Creation "science" do not count.

To see how severe the anomalies are, let us look at one example in more detail. Fossils of teleostean fishes (this class comprises just about all contemporary types from sardines to swordfish) are found only from late Triassic times (roughly 200 million years ago), and they show increasing abundance in the fossil record. Now recall that a leading principle of Flood Geology is that "animals living at the lowest elevations would tend to be buried at the lowest elevations" (Morris 1974a, 118). Overlooking such niceties as the fact that some teleosteans are deep-sea fish, let us ask what accounts for their success in resisting the Flood. Were they hydraulically special, less "streamlined" than other fish? No, as a group, there is more variation of shape *within* the teleosteans than there is between teleosteans and the groups of fish that were buried beneath them. So, perhaps the answer is that they found room at the top because they were upwardly mobile? But this explanation loses its attractiveness when we realize that the teleosts "range from speedy swimmmers to slow swimmers to almost sedentary forms, from dwellers in the open ocean to bottom-living types to lake and river fishes" (Colbert 1980, 57). Yet all these lucky teleostean fishes managed to resist the flood waters for a long time, while large numbers of speedy fish are buried beneath them.

By considering this one example, I hope to have explained what lies behind my charge that Flood Geology faces serious anomalies. But my principal purpose is to illustrate the impotence of the idea that worrisome details can be written off as "exceptional cases." Were *all* the teleosts exceptional? Was there no single unlucky sardine, salmon, or swordfish who was buried in the early deposits? Is it enough to remind ourselves that there are bound to be exceptions "because of the cataclysmic nature of the phenomena"? The case of the teleosts is only one among many. Ground sloths, flying reptiles, whales, trilobites, and a host of other creatures prove similarly embarrassing. Fleeming Jenkin, where are you when we need you?

Writing in 1961, Whitcomb and Morris made it clear what their last resort would be if the difficult questions began to threaten: "It is because the Bible itself teaches us these things that we are fully justified in appealing to *the power of God*, whether or not He used means amenable to our scientific understanding, for the gathering of two of every kind of animal into the Ark and for the care and preservation of those animals in the Ark during the 371 days of the Flood" (Whitcomb and Morris 1961, 79—italics in original). Today, "scientific" Creationists have pledged themselves to argue on the scientific evidence. So this last refuge is—officially, at least—out of bounds.

To each according to his need

The second major biological problem-solving strategy offered by Creationism attempts to answer questions about adaptation, relationships among organisms, and biogeographical distribution. Like evolutionary theory, Creationism hopes to deploy historical narratives to answer such questions. However, the style is very different. According to Creationists, the "basic kinds" of organisms were all created in a single event. Since their creation, and, in particular, since the Flood, the "basic kinds" have given rise to some new forms. However, although some later descendants of the original stock of a "basic kind" may be different from the ancestral organisms, they still belong to the same "basic kind." Small-scale evolution is admitted. What is denied is that evolutionary modification ever transgresses the boundaries of "basic kinds."

I postpone until later sections the issues raised by the Creationist attempt to drive a wedge between small-scale and large-scale evolution and by their idea of "basic kind." For the moment, let us just consider how their view of the historical development of life might be applied to biological problems. The characteristics of organisms can be understood in one of two ways. A feature of a kind of organism may either be viewed as an unmodified characteristic found among the organisms in the initial stock of that kind; or it may be identified as arising through modification of the properties of the original stock. Plainly, the latter idea can be employed only when the modifications are relatively slight, since only small-scale evolution is to be allowed. The relationships among groups of animals are understood by identifying the "basic kinds." Those groups—various breeds of dogs, perhaps—that descend from the same "basic kind" are related. Groups that descend from different "basic kinds" are not related. Finally, the distribution of organisms is explained by recounting the story of how the animals that survived the Flood have subsequently dispersed.

Although these ideas receive passing treatment in Creationist writings (see Morris 1974a, 72–75; Whitcomb and Morris 1961, 80–88), there is virtually no attempt to provide detailed explanations of biological phenomena. Occasionally, as in one of the "technical monographs"— entitled *Our Amazing Circulatory System . . . By Chance or Creation?* (Clark 1976)—Creationists sing the praises of one small part of the organic world. As I shall now argue, they need to do more than exclaim "Design!" when moved by the more striking instances of organic adaptation. Unless there is a specified account, which can be used to reveal the order of nature, no explanation has been given.

The most serious attempt to match the problem-solving success of evolutionary theory is made by Morris. In a short section of *Scientific Creationism*, he outlines the ways in which Creationism will tackle questions of biochemical similarity among groups of organisms, similarity of morphology and behavior across different kinds, and phenomena of convergent evolution (Morris 1974a, 71–75). Much of the discussion consists in ignoring the main point. Thus Morris mentions recent discoveries of biochemical similarities among organisms. This is a striking new success for evolutionary theory. Animals that share a recent ancestor turn out to have proteins with similar structures. (For example, the α chains of globin molecules are identical in humans and chimpanzees; human α globin chains differ from those of horses by 18 amino acids, from those of carp by 68 amino acids.) Evolutionary theory provides clear explanations of the numerous relationships unearthed by molecular techniques. What does Morris have to say about this? Nothing relevant. He suggests—quite wrongly—that the significance of the new discoveries is that biochemical similarities align themselves with "more traditional" anatomical and morphological criteria. (There is indeed a correspondence, but that is hardly what is most important about the findings.) At this point, he announces another triumph for Creationism. But evolutionary theory has shown itself able to answer a whole family of questions; the course of the evolution of the organisms can be seen reflected in the structural modification of biologically significant molecules. (There is a brief exposition in chapter 9 of Dobzhansky, Ayala, Stebbins, and Valentine 1977.) Morris has not demonstrated that Creationism can answer such questions. (For example, why do humans and chimpanzees have a common α globin?) He has just waved his hands in another direction.

Hints of a way of tackling some biological problems occur briefly at the end of the section: "[The creation model] suggests an array of similarities and differences, so that similarities simply suggest similar purposes (e.g., both birds and bats needed to fly, so the Creator created wings for both of them). This concept would apply equally well to so-called convergent evolution and cases of mimicry. All were created as distinct kinds, with similar structures for similar purposes and different structures for different purposes" (Morris 1974a, 75). The same idea has already been suggested much earlier in the book, where Morris contrasts the Creationist and evolutionist strategies of explanation: "The evolutionary explanation must be in terms of random variational processes producing a naturalistic evolutionary chain all the way from particles to people. The creationist explanation will be in terms of primeval planning by a personal Creator and His implementation of that plan by special creation of all the basic entities of the cosmos, each with such structures and such behavior as to accomplish most effectively the purpose for which it was created" (Morris 1974a, 33). Elsewhere, citing Newton and Kepler, Morris alludes to the joy of discovering the beauty and pattern in nature, and so "thinking God's thoughts after Him" (Morris 1974a, 14).*

The basic idea is straightforward. When we recognize a characteristic of an organism as unmodified, the Creationist explanation of its presence will be to show how the feature manifests the Creator's design. The account of similarities among distinct "basic kinds" will identify the similar needs of organisms of those distinct kinds. Here are some sample "explanations." Bats have wings because the Creator endowed them with wings, and He did so because they need to fly. Chimps and humans have similar hemoglobins (and other biological molecules) because the Creator gave them similar molecules from the start, and He did so because their physiological requirements are similar. Some (perfectly palatable) butterflies mimic unpalatable butterflies of the same region because the Creator saw that they had to have some defense against predators. To each according to his need.

If one wants to believe in Creationism, the picture can easily lull critical faculties. Yet, if we think about it, it is bizarre. Surely we should not imagine the Creator contemplating a wingless bat, recognizing that it would be defective, and so equipping it with the wings it needs. Rather, if we take the idea of a single creative event seriously, we must view it as the origination of an entire system of kinds of organisms, *whose needs themselves arise in large measure from the character of the system.* Why were bats created at all? Why were any defenses against predation needed? Why did the Creator form this system of organisms, with their interrelated needs, needs that are met in such diverse and complicated ways?

Invocation of the word "design," or the passing reference to the satisfaction of "need," explains nothing. The needs are not given in advance of the design of structures to accommodate them, but are themselves encompassed in the design. Nor do we achieve any understanding of the adaptations and relationships of organisms until

---

*The contemplation of nature can give rise to some curious reflections. There is a famous (possibly apocryphal) story about the great biologist J. B. S. Haldane. At a major British public occasion, Haldane was sitting next to an Anglican bishop, who asked him what biology had shown him about the designs and predilections of the Creator. Haldane is supposed to have replied, "An inordinate fondness for beetles."

we see, at least in outline, what the Grand Plan of Creation might have been. This point has been clear at least since the seventeenth century. At the beginning of the *Discourse on Metaphysics*, Leibniz gave a beautiful exposition of it. He recognized that unless there are independent criteria of design, then praise of the Creator's design is worthless: "In saying, therefore, that things are not good according to any standard of goodness, but simply by the will of God, it seems to me that one destroys, without realizing it, all the love of God and all his glory; for why praise him for what he has done, if he would be equally praiseworthy in doing the contrary?" (Leibniz 1686/1979, 4–5). For Leibniz, to invoke "design" without saying what counts as good design is not only vacuous but blasphemous. Later in the same work, Leibniz developed the theme with a striking analogy. *Any* world can be conceived as regular ("designed") just as *any* array of points can be joined by a curve with some algebraic formula.

Why are contemporary Creationists silent about the Design? Because things did not go so well for their predecessors who tried to show how each kind of organism had been separately created with a special design. They found it hard to reconcile the observed features of some organisms with the attributes of the Creator. Contemporary Creationists have learned from these heroic—but fruitless—efforts.

So we encounter the strategy exemplified by Morris: Talk generally about design, pattern, purpose, and beauty in nature. There are many examples of adaptations that can be used—the wings of bats or "the amazing circulatory system," for example. But what happens if we press some more difficult cases? Well, if there seems to be no design or purpose to a feature (and if its presence cannot be understood as a modification of ancestral characters), one can always point out that some parts of the Creator's plan may be too vast for human understanding. *We* do not see what the design is, but there *is* design, nonetheless.

Since no plan of design has been specified, Creationists have available another all-purpose escape clause. But it is precisely this feature of Creation "science" that impugns its scientific credentials. To mumble that "the ways of the Creator are many and mysterious" may excuse one from identifying design in unlikely places. It is not to do science.

To provide scientific explanations, a Creationist would have to identify the plan implemented in the Creation. The trouble is that there are countless examples of properties of organisms that are hard to integrate into a coherent theory of design. There are two main types of difficulty, stemming from the frequent tinkerings of evolution and the equally common nastiness of nature. Let us begin with evolutionary tinkering. Structures already present are modified to answer to the organism's current needs. The result may be clumsy and inefficient, but it gets the job done. A beautiful example is the case of the Panda's thumb (Gould 1980). Although they belong to the order Carnivora, giant pandas subsist on a diet of bamboo. In adapting to this diet, they needed a means to grasp the shoots. Like other carnivores, they lack an opposable thumb. Instead, a bone in the wrist has become extended to serve as part of a device for grasping. It does not work well. Any competent engineer who wanted to design a giant panda could have done better. But it works well enough.

It is easy to multiply examples. Orchids have evolved complicated structures that discourage self-fertilization. These baroque contraptions are readily understood if we understand them as built out of the means available. Ruminants have acquired very complicated stomachs and a special digestive routine. These characteristics have enabled them to break down the cellulose layers that encase valuable proteins in many grasses. Their inner life could have been so much simpler had they been given the right enzymes from the start.

The second class of cases covers those in which, to put it bluntly, nature's ways are rather repulsive. There is nothing intrinsically beautiful about the scavenging of vultures, the copulatory behavior of the female praying mantis (who tries to bite off the head of the "lucky" male), or the ways in which some insects paralyze their prey. Let me describe one example in more detail. Some animals practice *coprophagy*. They produce feces that they eat. Rabbits, for example, devour their morning droppings. From an evolutionary perspective, the phenomenon is understood. Rabbits have solved the problem of breaking down cellulose by secreting bacteria toward the end of their intestinal tracts. Since the cellulose breakdown occurs at the end of the tract, much valuable protein and many valuable bacteria are liable to be lost in the feces. Hence the morning feces are eaten, the protein is metabolized, and the supply of bacteria is kept up. Creationists ought to find such phenomena puzzling. Surely an all-powerful, all-loving Creator, who *separately designed* each kind of living thing, could have found some less repugnant (and, I might add, more efficient) way to get the job done. (These examples are, of course, far less problematic for those who believe simply that the Creator set the universe in motion billions of years ago and that contemporary organisms are the latest product of the laws and conditions instituted in that original creative event.)

So far, I have concentrated on Creationist resources for answering questions about the characteristics of organisms and the similarities and differences among kinds of organisms. Let us now take a brief look at Creationist biogeography. Discussions of the distribution of animals are not extensive, but the following passage lays down the ground rules of the enterprise: "If the Flood was geographically universal, then all the air-breathers of the animal kingdom which were not in the Ark perished; and present-day animal distribution must be explained on the basis of migrations from the mountains of Ararat" (Whitcomb and Morris 1961, 79). One's first response is surely to ask: Why only one Ark? Why Ararat? (Why not New Jersey?) Of course, we know the answers to these questions. But what *scientific* evidence is there for supposing that there was just one vehicle for preserving land animals during the Deluge and that the subsequent radiation began from Mount Ararat? Creationists tie their hands behind their backs when they approach problems of biogeography with such gratuitous assumptions. There are obvious difficulties posed by the existence of peculiar groups of animals in particular places. The most striking example is the presence of marsupials as the dominant mammals of Australia. Given that all the land animals reemerged at the same place at the same time, why did Australia become a stronghold for marsupial mammals?

Whitcomb and Morris consider precisely this question. Much of their discussion is directed against claims made by one of their evangelical rivals, a geologist who advocated only a "local flood." However, they do indicate the main lines of their answer. In essence, they propose to accelerate the migration of organisms described in a standard evolutionary account. Here is the standard explanation of how the marsupials came to dominate Australia.

One central hypothesis is *placental chauvinism*: Marsupials are competitively inferior to the recent eutherian (placental) mammals. This hypothesis is confirmed by evidence of the consequences of introducing eutherian competitors into marsupial populations. It is usually suggested that the marsupials arose in North America about 130 million years ago and that they were able to compete successfully with the *early* eutherian mammals. The marsupials radiated extensively, established themselves in South America, and crossed over to Australia by way of Antarctica. (Australia, Antarctica, and South America became separated about 70 million years ago.) Their eutherian contemporaries did not reach Australia, so that the marsupials were able to diversify and attain their modern forms without competition from eutherians. Other marsupial strongholds (for example, South America) became vulnerable when new continental connections (the Isthmus of Panama) made it possible for the highly successful *recent* eutherians to invade.

But Australia was sufficiently isolated, and a rich marsupial fauna developed there.

What Creationists propose to do is to squash something like this sequence of events into less than 100 centuries. Here, then, is the scenario. Noah's Ark lands on Mount Ararat. Out come the animals. They begin to compete for resources. Because they are inferior competitors, the poor marsupials are forced to disperse ever more widely. Spreading southward, they eventually manage to reach Australia. By the time the placentals have given chase, the land connection with Australia is severed. The marsupials are safe in their stronghold.

This is an exciting story, worthy of the best cowboy tradition. The trouble is that it has the marsupials arising in the wrong place, going by the wrong route, and competing with the wrong animals. Apart from that, the pace is just a bit too quick. If Creationists are going to explain fossil findings that, by their own lights, are post-Flood, they had better suppose that the marsupials reached Australia by travelling through Europe, North America, and South America. If they are going to insist that *contemporary* kinds of eutherian mammals emerged from the Ark, then they will have to explain why competition was not so severe that the marsupials were completely vanquished. Waiving these difficulties, let us consider the rate of the migration. When we think of marsupials, we naturally think of kangaroos—so we have the vision of successive generations of kangaroos hopping toward Australia. But kangaroos are relatively speedy. Some marsupials—wombats, koalas, and marsupial moles, for example—move very slowly. Koalas are sedentary animals, and it is difficult to coax them out of the eucalyptus trees on which they feed. Wombats, like marsupial moles, construct elaborate burrows, in which they spend their time and carry on their social relations. The idea of *any* of these animals engaging in a hectic dash around the globe is patently absurd. (On the evolutionary account, of course, they are all descendants of ancestral marsupials who had millions of years to reach their destination.)

Next we must face the question of why all the lucky refugees were *marsupials*. Surely, large numbers of animals would have found it prudent to disperse widely from the Ark. Why is it that the marsupials, almost *alone* among the mammals, were able to find the land connection to Australia and scurry across before other mammals in need of *Lebensraum* could catch up with them? And what about the conveniently disappearing land connection itself? Creationists seem to assume very rapid movements of land masses. Unlike orthodox geologists, who have independent evidence for slow separation of the continents, they maintain that, in a matter of centuries, a land connection that would support a full-scale exodus of marsupials presented an insuperable barrier to the pursuing eutherians. Indeed, if the marsupials were really *driven* across by eutherian competition, then we would expect the competition to be snapping at their heels—otherwise would not the wombat have stopped to dig a burrow, the koala have settled in a convenient tree? In that case, the bridge would have to be cut *very* quickly.

Once again, when the Creationist story is pressed for details, anomalies appear in droves. Moreover, the final issue should remind us again of Fleeming Jenkin's complaint against Darwin: What are the rules of the Creationist game? What constraints govern hypotheses about past land connections? Since the Creationists have foresworn the apparatus of modern geology, their claims about the past relations of land masses seem invulnerable to independent checks. No worries about mechanisms for rapid land subsidence need perturb them, for they may always appeal to the after effects of the great cataclysm. Anything goes.

When Whitcomb and Morris wrote *The Genesis Flood* in 1961, Creationist strategy was somewhat different from that currently in vogue. Those were halcyon days, when Creationists did not mind admitting their reliance on unfathomable supernatural mechanisms. Perhaps they even hoped that a version of Creationism, explicitly based on the Genesis account, might find its place in science education. The following passage is far less guarded than more recent statements:

The more we study the fascinating story of animal distribution around the earth, the more convinced we have become that this vast river of variegated life forms, moving ever outward from the Asiatic mainland, across the continents and seas, has not been a chance and haphazard phenomenon. Instead, *we see the hand of God guiding and directing these creatures in ways that man, with all his ingenuity, has never been able to fathom*, in order that the great commission to the postdiluvian animal kingdom might be carried out, and "that they may breed abundantly in the earth, and be fruitful, and multiply upon the earth" (Gen. 8 : 17). (Whitcomb and Morris 1961, 86—italics mine)

There is the all-purpose escape clause. If the way in which the animals might have managed to leave the Ark and distribute themselves around the globe boggles your mind, do not tax yourself. They were guided, directed in ways that we are not able to understand.

Morris's subsequent writings take a different line about biogeographical questions. The subject is not discussed. Hence it is impossible to be sure that current Creationists would invoke the actions of the

Creator to help out when the going gets tough. Nevertheless, this is one more instance of the phenomenon that we have seen repeatedly. The alleged rival to evolutionary theory provides no definite problem-solving strategies that can be applied to give detailed answers to specific questions.

Debates about macroevolution

Although we have examined the biological problem-solving strategies that are native to Creationism, we have not exhausted the Creationists' repertoire. They believe that they can take over, unmodified, much of the problem-solving apparatus of contemporary evolutionary theory. This surprising suggestion rests on the claim that there is an important distinction between small-scale evolution (microevolution) and large-scale evolution (macroevolution). Creationists are perfectly willing to allow for descent with modification, even to suppose that a single species can split into two descendant species. What they deny is that one *kind* of organism can evolve from another *kind*.

The distinction between evolution within kinds and evolution across kinds is made again and again in Creationist writings:

Breeding experiments with plants and animals afford extensive data, usable in support of the Limited Evolution Model; and, to avoid equivocation of terms the phenomena involved might just as well be called "genetic variation." All known, observable changes of living things are always within recognizable limits of variation of major groups of plants and animals. (Moore 1974, 3)

It is normal variation of this sort, unfortunately, which is still commonly offered as evidence of present-day evolution. The classic example of the *peppered moth* of England, "evolving" from a dominant light coloration to a dominant dark coloration, as the tree trunks grew darker with pollutants during the advancing industrial revolution, is the best case in point. This was not evolution in the true sense at all but only variation. Natural selection is a conservative force, operating to keep kinds from becoming extinct when the environment changes. (Morris 1974a, 51)

In our discussion of evolution, therefore, we are *not* referring to the possible origin of the variations within the dog kind. We are referring to the alleged origin of the dog kind and cat kind from a common ancestor. We are *not* referring to the origin of the finches within *Geospiza*, *Camarhynchus*, and *Cactospiza*. We *are* referring to the origin of these finches and, say, the herons, from a common ancestor, and their ultimate origin from an ancestral reptile. (Gish 1979, 38)

Hence, if you ask Creationists how they propose to solve some of the classic problems of evolutionary biology—What accounts for the characteristics of the Galapagos finches? Why do melanic forms become dominant in moth populations after the Industrial Revolution?—they have a simple reply. Their solution is the standard evolutionary solution. They maintain that Creationism retains all the real problem-solving power of evolutionary theory; all that is jettisoned are some wild speculations.

Many evolutionary biologists would find this suggestion absurd. They would argue that there is no basis for a distinction between the processes of microevolution and those of macroevolution. As I noted in chapter 1, contemporary neo-Darwinian orthodoxy is *evolutionary gradualism*. Speciation is regarded as a slow process in which many small genetic changes accumulate to result in the reproductive isolation of two populations. Population genetics studies these small genetic changes over short periods of time. Gradualists contend that one cannot admit the findings of population genetics without conceding the likelihood of major evolutionary changes by similar mechanisms. For large-scale evolution differs only in degree from the small modifications that are observed and studied. It is simply a matter of more changes, extended over a longer time.

Gradualist orthodoxy has been challenged, and Creationists have seized upon the words of the challengers. There are two independent challenges, that are frequently conflated by Creationists (and sometimes by biologists). The first challenge concerns the *tempo* of evolution. In a seminal paper, Niles Eldredge and Stephen Gould offered an alternative to the gradualist account of the pace of evolutionary change. Their proposal, the *punctuated-equilibrium model*, begins by building upon Mayr's theory of geographic speciation (Mayr 1963; Mayr 1970). Instead of thinking of species as evolving slowly and continually, Eldredge and Gould suggested that, in small isolated populations, subjected perhaps to special environmental demands, evolutionary change may be very rapid (in geological time). Stasis, the absence of change, is the norm for a species. Its central population is expected to persist relatively unmodified for millions of years. Detached from the central population, small peripheral groups can undergo morphological change very quickly, so that a period of isolation of a few thousand years may produce a population that is reproductively isolated and morphologically distinct from the parental stock. Evolution is not a slow continuous process but a jerky affair. There are long pauses when nothing happens, interspersed with frenzied bursts of activity.

This first challenge to gradualism offers a different account of the "geometry" of evolution. The kinds of "trees" that evolutionary theorists have been drawing throughout most of the past 125 years give way to new pictures (see figure 5.1). Now this is not simply a disagreement about how to construct diagrams. The models will take quite different views of the fossil findings. For example, imagine that the fossil record of a type of snail shows a trend toward increased height of the shell, so that, in *both* evolutionary models, it is concluded that there was a succession of species, each with a greater average shell height than its predecessor. This situation, together with interpretations that might be favored by gradualists and punctuated-equilibrium theorists, respectively, is depicted in figure 5.2. Notice that the two types of interpretation lead to very different expectations about fossil findings. Gradualists suppose that if we do not find a smooth sequence of intermediate fossil forms, then we must lay the blame on the difficulty of being fossilized. By contrast, punctuated-equilibrium theorists describe the rapid variation of small, isolated populations who would be expected not to have left traces of themselves behind.

So far, the attack on gradualism has little bearing on the issue of whether Creationists can coopt parts of evolutionary biology—specifically standard population genetics—without committing themselves to the possibility of the sorts of changes (evolution across "kinds") that they take to be impossible. The clash of ideas about the tempo of evolution does not yet amount to a conflict about the genetic mechanisms of evolutionary change. Hence it does not support the conclusion that the mechanisms of large-scale change are different from those that have been studied by population geneticists. Nevertheless, the debate already provides ammunition for Creationists to use.

Nobody has used the ammunition with more gusto than Duane Gish. The third edition of *Evolution? The Fossils Say No!* contains an extended discussion of current debates within evolutionary theory, designed to play defenders of the punctuated-equilibrium model against the gradualists. Here, for example, is Gish's discussion of a popular article by Gould ("The Return of the Hopeful Monster," reprinted in Gould 1980): "Somewhat later in the same article Gould says: 'All paleontologists know that the fossil record contains precious little in the way of intermediate forms; transitions between major groups are characteristically abrupt' ". (Gish 1979, 172). This citation and others that are similar, should not mislead us into thinking that Creationist claims about the fossil record are upheld by reputable paleontologists. Punctuated-equilibrium theorists do not deny that, *in one sense*, there

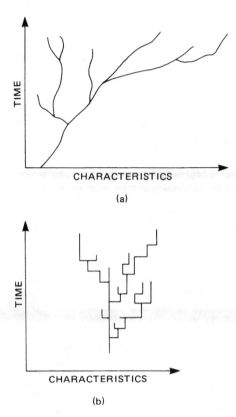

(a)

(b)

Figure 5.1
The nature of evolutionary change as viewed by the (a) gradualist and the
(b) punctuated-equilibrium models. The horizontal axis represents a measurement
of the characteristics of the organisms in question. Obviously, a more complete
representation, taking into account many characteristics separately, would require
a multidimensional space (that is, more than two axes).

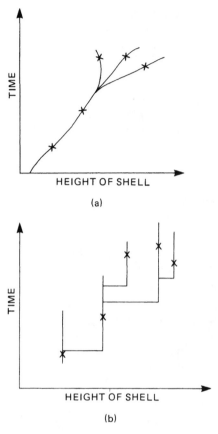

(a)

(b)

Figure 5.2
The nature of evolutionary change as viewed by the (a) gradualist and the
(b) punctuated-equilibrium models. The crosses represent fossil findings. Note that
the example is purely illustrative. I certainly do not mean to suggest that these
are the *only* pictures that might be drawn, or that one is entitled to draw *any*
picture, given just these data.

are transitional forms between classes of organisms. They are happy to acknowledge that *Archeopteryx* is a transitional form between reptiles and birds, since it is an animal with some characteristics of both classes. (I shall consider in the next chapter Gish's most recent attempt to muddy the waters on this point.) They are equally clear that the fossil record shows a sequence of therapsids becoming ever more mammallike. What is denied is that there are *smooth* and *gradual* transitions among *species*. Focusing on the relatively small differences between related species, punctuated-equilibrium theorists point out that we do not find these differences bridged by a continuous sequence of intermediates. However, to deny the existence of intermediates on a fine scale is not to suggest that we never encounter intermediates between large groups of organisms, creatures that exhibit fully developed features of the later group while retaining some ancestral characteristics.

Although defenders of the punctuated-equilibrium model often support their thesis by appeal to the fossil record, their point is radically different from the claims of the Creationists. However, there is a second challenge to gradualism that Creationists try to adapt to their purposes. Some contemporary biologists have questioned gradualism as an account of the *mode* of evolution. They argue for either or both of two different claims: (1) Major morphological changes need not be produced by successive substitution of alleles with small phenotypic effects, but by *macromutations*, small genetic changes producing large phenotypic alterations; (2) Reproductive isolation may result from a small number of genetic changes at important loci. I want to emphasize that neither of these claims is a necessary consequence of the punctuated-equilibrium model alone. Further argument is required to show that occasional bursts of evolutionary change (which are rapid on the scale of *geological time*, producing speciation in a few thousand years) require macromutations.

Some biologists, notably Gould, think that the further arguments can be given and that gradualists are wrong about both the tempo and the mode of evolution. Gould denies that the well-understood cases of allelic replacement in fruit flies or peppered moths provide a basis for extrapolation. He maintains that large-scale morphological shifts need not result from a succession of genetic changes, each producing a small phenotypic effect. In this context, he has had words of praise for the unorthodox evolutionary theorist Richard Goldschmidt, whose suggestion that large evolutionary changes take place through the emergence of "hopeful monsters" has frequently been ridiculed.

Gish pounces on this controversy, and on Gould's reference to Goldschmidt:

Goldschmidt termed his mechanism the "hopeful monster" mechanism. He proposed, for instance, that one time a reptile laid an egg and a bird was hatched from the egg! All major gaps in the fossil record were accounted for, according to Goldschmidt, by similar events—something laid an egg, and something else got born! Neo-Darwinists prefer to believe that Goldschmidt is the one who laid the egg, maintaining that there is not a shred of evidence to support his "hopeful monster" mechanism. Goldschmidt insists just as strongly that there is no evidence for the postulated neo-Darwinian mechanism (major transformations by the accumulation of micromutations). Creationists agree with both the neo-Darwinists and Goldschmidt—there is no evidence for *either* type of evolution. (Gish 1979, 170)

The first point to realize is that there are significant differences between the standard interpretation of Goldschmidt's ideas and the proposals about macromutations that are currently being discussed. So far as I know, nobody is currently defending the idea that the birds evolved through the emergence of a single individual bird from a reptilian egg. The principal suggestion has been that some mutations—perhaps mutations in regulatory genes—might produce large effects by altering the timing of developmental events. By modifying the pattern of development, they could produce organisms with a different form from that of the parents.

I shall not investigate the issue of exactly who is entitled to wear the badge of neo-Darwinism. (Some biologists believe that the alleged challenge to neo-Darwinism has been given more publicity than it deserves. They maintain that, insofar as there is a defensible notion of macromutation, it is already accommodated by orthodox neo-Darwinism. For simplicity's sake, I assume that there is a genuine dispute and consider its implications for Creation "science.") In essence, there are two opposed positions about the mode of evolution: (1) Some large-scale evolutionary changes have involved macromutations; the experiments of classical population genetics have not revealed any small genetic changes resulting in advantageous morphological discontinuities; classical population genetics is therefore not a complete account of the genetics of the evolutionary process: (Population geneticists have not yet studied all the interesting evolutionary phenomena.) (2) All evolutionary changes involve mutations that are no different in kind from those studied in the experiments of classical population genetics. Gish wants to play these positions off against one another. He selects from (1) the point that some evolutionary changes require (or would require) advantageous macromutations, and draws from (2) the conclusion that we have no reason to believe that advantageous macromutations ever occur. The result of this mixture is an attempt to suggest that the changes in question could not have taken place.

This reasoning distorts the situation. There is an ongoing debate about the *mechanism* of evolution. That debate does not touch the *fact* of evolution. Evolutionary theorists often agree that two particular species evolved from a common ancestor, even though they disagree about how the evolution occurred. One theorist may suggest that the process involved a succession of genetic changes with small phenotypic effect. The other may insist that the transition required mutations affecting developmental patterns, so that small genetic alterations were amplified in the phenotype. It is illegitimate to deploy the arguments offered by both sides, arguments that criticize the rival accounts of the *process* of evolution, to cast doubt on the *existence* of evolution. *For we can know that a species is related to an ancestral population by evolutionary descent, even though the details of the transition are controversial.* In the last three sentences of the passage just quoted, Gish slides from claims that there is no evidence for mechanisms of evolution to the conclusion that there is no evidence for evolution. This is simply to ignore all the reasons that scientists may have for recognizing an evolutionary relationship, such as intricate similarities in morphology. To use a case that I described in the last chapter, we know that the mammals evolved from the therapsids because we can recognize similarities and trends in the fossil record. But we do not know everything we would like to know about this case. There is continuing debate about how the early mammals are interrelated and how rapidly important transitions occurred.

The suggestion that macroevolution should be divorced from micro-evolution provides Creationists only with a debating point. It allows Creationists to say that there are some evolutionary theorists who distinguish the mechanisms studied in classical population genetics from those they take to be involved in large-scale evolutionary change (or, more exactly, in some cases of large-scale evolutionary change). But this is not to suppose that the distinction drawn by heterodox evolutionists is that favored by the Creationists. (The vagaries of the Creationist notion of kind will be considered in the next section.) More fundamentally, to query traditional ideas about the mechanisms of large-scale evolutionary change is not to deny the existence of episodes of large-scale evolutionary change, or to cast doubt on the evidence that shows us that those episodes took place. The challenge to neo-Darwinism calls for a broader view of the genetics of the evolutionary process. It continues to emphasize the compelling reasons that lead us to believe that organisms are evolutionarily linked. In other words, the debate neither questions the existence of evolutionary relationships nor rejects the evidence (drawn from morphological similarities and so forth) on which claims about evolutionary relations rest; what is in dispute is the explanation of how evolutionary changes occurred.

It is all too easy for internal scientific disputes to confuse outsiders into thinking that the foundations of the discipline are being undermined. A very recent newspaper article, under the headline "Darwin Theory Is Challenged," began with the following sentence: "Humans did not evolve gradually from ape-like ancestors but moved up the evolutionary scale in quantum leaps set off by the environment, a Johns Hopkins University scientist [Steven Stanley] says in the most recent challenge to Darwin's theory of evolution" (*New Haven Register* November 12, 1981). Many people, reading the headline and the first sentence, passing over the important word 'gradually' and not understanding the hazy formulation about "quantum leaps," will misunderstand the proposal and assume that what is being denied is the claim that humans evolved from apelike ancestors. Fortunately, the rest of the article does better, but I am not confident that all those who read it saw that the issue was the mode of evolution, rather than the existence of evolution. The debate about the tempo and mode of evolution—which has generated some of the most fruitful discussion in the recent history of evolutionary theory—can easily be misinterpreted as scientific support for the Creationist cause.

What's in a kind?

We have seen that Creationists would like to coopt explanations from evolutionary theory and so account for certain classic cases of modification or diversification of a group of organisms. We have seen how they try to use a debate within evolutionary theory to their advantage, and why this tactic is illegitimate. Let us turn now to a different point. Can Creationists find a notion of kind that will meet their needs? Can they characterize kinds sufficiently broadly to use the idea of evolution within a kind to resolve biological questions? At the same time, can they make kinds sufficiently narrow to accommodate their main thesis that no significant evolution occurs?

There is some uncertainty among Creationists about how the term *kind* is to be defined. Hiebert introduces the expression *basic kind* with a parenthetical remark, asserting that basic kinds correspond "roughly to the arbitrary category of species" (Hiebert 1979, 37). Morris uses "interfertility" as a criterion for assigning organisms to the same kind (Morris 1974a, 180). Gish offers a more elaborate discussion: "We must here attempt to define what we mean by a basic kind. A basic animal or plant kind would include all animals or plants which were truly derived from a single stock. In present-day terms, it would be said that they have shared a common gene pool. All humans, for example, are within a single basic kind, *Homo sapiens*. In this case, the basic kind is a single species. In other cases, the basic kind may be at the genus level" (Gish 1979, 34). Gish goes on to cite the species of the coyote, the wolf, the jackal, and the dog as making up a basic kind "since they are all interfertile and produce offspring" (Gish 1979, 36). Wysong is even more informative. He begins by differentiating Creationist taxonomy from standard scientific taxonomy: "It is important to understand that the species classification was not built entirely upon the criteria used to delineate creation kinds, i.e. interfertility" (Wysong 1976, 59).

The trouble is that Creationists have to "expand" the interfertility criterion. They do not want to claim that *Drosophila* species that are not interfertile belong to different kinds. (Why not? Because in this case there would be convincing evidence for evolution across kinds.) Hence there is an appeal to morphology: "If two organisms breed, even though it is infrequent, they are of the same kind; if they don't breed but are clearly of the same morphological type, they are of the same kind by the logic of the axiom which states two things equal to the same thing are equal to each other" (Wysong 1976, 60). Wysong's application of the logic of identity is grotesque. However, disregarding this point, the main issue concerns what criterion of "same morphological type" is to be employed. Wysong is ready for such worries: "I can see some difficulties with the precise details of this definition but it can nevertheless serve us well here as a working hypothesis" (Wysong 1976, 60). Wysong is wrong. If Creation "science" is to be taken seriously, so flexible a definition will not do.

It is easy to see why Creationists want to leave matters vague. As Gish candidly remarks, basic kinds are those that descend from an original stock. This will not do as a definition, of course, since the whole point of dispute concerns which organisms descend from the same original stock. Yet Gish's comment serves to remind us that the point of the enterprise is to identify the boundaries across which evolution is not to occur. What are the possibilities?

One possible line is suggested by Hiebert: Basic kinds are species. But this will lead to trouble because evolutionary theorists can produce clear and well-understood examples of transspecific evolution. Darwin's finches are different species, and the genus *Drosophila* includes numerous species with well-charted evolutionary relationships. Rather than take on these difficult cases, Creationists beat a retreat.

Now it would clearly be counterproductive to identify basic kinds with some standard taxonomic category above the species level, to propose, for example, that basic kinds are genera or families. Creationists want to claim that some species, like *Homo sapiens*, are basic kinds. So they are forced into crude gerrymandering. What is needed is a concept of kind that cuts across the standard taxonomic hierarchy. The main suggestion offered is that organisms belonging to the same kind are interfertile. What is intended here is not the modern concept of species, according to which two populations constitute distinct species if they do not interbreed in nature (that is, if they are reproductively isolated from one another). It is well known that many groups of organisms that do not interbreed in nature are interfertile in the sense that they can be induced to mate, and produce offspring, in captivity. (Conversely, some conspecific organisms do not take kindly to mating in captivity; witness the unfortunate Giant Panda.) Because of the difference between mating in the wild and mating in captivity, the interfertility criterion does not reduce to the standard definition of species. Hence interfertility looks like a promising criterion for basic kinds.

The trouble is that this will not quite do. On the one hand, appeal to the interfertility criterion will mark as distinct kinds species that have a clear evolutionary relationship. There are species of *Drosophila*, for example, whose chromosomes are clearly derived from a common ancestor, but that divide up the chromosomal material differently. Most of these species are unable to produce fertile hybrids. Some are unable to produce any hybrids at all. Wysong is aware of such cases, and broadens his definition of *kind* to circumvent them. But there is a problem from the other direction. Appeal to interfertility allows lions and tigers to belong to the same kind. If this usage of *kind* is strictly observed, are we to assume that some common ancestor of lions and tigers was orginally created, and that lions and tigers have since evolved from that common ancestor? Or did the Creator originally make several "varieties" of some kinds? For example, in creating some feline kind, were lions and tigers made separately as "individual varieties?" If so, what was the rationale for favoring some kinds with different "varieties"? If not, the Creationist must assume some very rapid events of diversification (the emergence of lions and tigers from a common feline ancestor; of dogs, wolves, and jackals from a common canine ancestor).

Wysong's conception of basic kinds is very vague. Since he makes no attempts to define the level of morphological similarity that suffices to include two organisms within the same kind (even though they are not interfertile), he is able to adjust his taxonomy as he sees fit. Should he assume that all rodents constitute a basic kind? All snails? All fish? Given an appropriate conception of morphological similarity, each of these groups could be identified as a basic kind. However, it is evident that so broad a view of kind would raise problems for the Creationist idea that the basic kinds were originally created and have undergone *minor* modifications ever since. In fact, the strategy is clear. The idea of morphological similarity will be adjusted to allow Creationists to make any classifications they want.

Yet it is not clear that the retreat into vagueness helps. The initial motivation for developing the concept of kind was to apply evolutionary problem-solving strategies on a limited scale. Creationists hope to borrow the usual evolutionary account of the distribution of finches in the Galapagos, according to which ancestral populations reached the islands from Ecuador and became diversified. But even though they broaden the notion of basic kind, there will remain some cases of evolutionary explanation that they cannot absorb. At one, relatively unguarded moment, Gish announces that "the gibbons, orangutans, chimpanzees, and gorillas would each be included in a different basic kind" (Gish 1979, 37). This decision deprives the Creationist of any ability to borrow evolutionary explanations of the many great similarities among these primates. From the molecular biologists' discoveries of close kinship in molecular structures of proteins to the well-established similarities of anatomical structure, there is a host of details that cry out for explanation in terms of common ancestry. It seems that Creationists will strain at finches and fruit flies only to swallow the apes.

What we have encountered here is the same indefiniteness that appears in the appeals to Flood Geology and the Creator's design. But, in this case, there is something new. Creationists are sufficiently impressed with some evolutionary explanations that they would like to borrow them for their own use. Having once recognized the power of common ancestry to explain the chromosomal similarities among species of fruit flies or the kinship of the Galapagos finches, they draw the line at using the same type of explanation to account for other powerful similarities (such as the kinship of the gibbons and the great apes). Now it is the essence of unification to use the same pattern of reasoning to solve diverse problems. Yet, having applied the pattern of evolutionary explanation where it suits their purposes to do so,

Creationists refrain from applying it elsewhere, *without offering any coherent account of the distinction between the cases.* They offer us a hodge-podge, picking and choosing from evolutionary theory, not by applying some principled criterion but by tailoring the concept of kind to suit the needs of the moment.

One final point. Gish's list of basic kinds identifies bats, an entire diverse order, containing more than 850 species, as a single kind. (The variation among bats is enormous; there is a wide range of dietary habits, sizes and flight pattern; some bats see, others do not.) Yet in the primate order, the family Pongidae (great apes and gibbons) is divided into four separate kinds. Why? Well, of course, this is too close to home. The morphological similarities among the pongids are extensive, but to admit these as important evidence of common ancestry would be to flirt with disaster. Humans are just not that different from the great apes. "If the same standards used in the classification of other mammalian orders were applied to the primates, man would be included with the gibbons and great apes in a single family" (Vaughan 1978, 146). Hence differences that are dismissed in the cases of bats and cats (and, no doubt, bees and trees, whales and snails) suddenly become important when the major point of Creationism might be threatened.

Blind dates

I now turn to the last gasp of the Creationists' "scientific" defense of their theory. We have looked at a "theory" that has no detailed problem solutions to its credit (except those it borrows from its rival), that has no clearly defined problem-solving strategies, that encounters anomalies whenever it becomes at all definite, but that typically relapses into vagueness whenever clear-cut refutations threaten. Why should we take this "theory" to be worthy of any consideration? Creationists can claim that there is one definite difference between Creation "science" and evolutionary theory, one point that Creationists clearly and explicitly make. Further, they would claim that the evidence favors their view. The point in question concerns the age of the earth. Creationists maintain that, contrary to the teaching of evolution, the earth is relatively young, merely a few thousand years old.

Before we take a look at arguments for a young earth, it is worth noting that, even here, there are moments of characteristic Creationist diffidence. As he prepares to consider the evidence for claims about the age of the earth, Morris cannot refrain from hedging his bets: "As a matter of fact, the creation model does not, in its basic form, *require* a short time scale. It merely assumes a period of special creation sometime in the past, without necessarily stating when that was" (Morris 1974a, 136). So qualified, Creationism need not fear the results of scientific dating procedures. Equally, no evidence for a young earth can redound to its credit.

The standard scientific estimate is that the universe is about 15 billion years old, the earth about 4.5 billion years old. It is important to recognize from the start that there are independent procedures for obtaining each of these estimates, and that the procedures yield ranges of values that overlap. In the case of the universe, estimates can be obtained from astronomical methods or considerations of nuclear reactions. Astrophysicists can measure the rate at which galaxies are receding and use these measurements to compute the time needed for the universe to expand to its present size. A second, independent, astronomical method is to use standard techniques to measure some parameters of stars (mass, luminosity, compositon, and surface temperature), from which a well-confirmed theory of the life histories of stars enables physicists to compute their ages. Finally, considerations of radioactive decay make it possible to calculate the time at which certain heavy elements were formed. These techniques are somewhat similar to the radiometric methods of dating rocks, which I shall consider in a little more detail. (For an excellent overview of the various ways of assigning an age to the universe, and an exposition of the radioactive decay method, see Schramm 1974.)

Although the clear consensus of physical techniques is that the universe is billions of years old, and although this result controverts the claims of at least some contemporary Creationists, the principal Creationist attack has been directed against the standard geological claim that the earth is about 4.5 billion years old. Two kinds of arguments are offered. In the first place, Creationists argue that methods of radiometric dating employ false assumptions. They continue by using special techniques of their own to assign to the earth an age of a few thousand years. Excellent and exhaustive explanations of the errors in Creationist arguments about dating methods have been given by Stephen Brush (1982) and Brent Dalrymple (forthcoming). My aim in the following brief discussion is simply to hit the high spots.

The basic idea behind radioactive dating is very simple. If a radioactive isotope (the *parent element*) was originally present in a rock at the time of its formation, then that isotope would give rise, by radioactive decay, to decay products (*daughter elements*). The phenomenon of radioactive decay is well understood by nuclear physicists. It is governed by the following equation:

$$N_t = N_0 e^{-\lambda t}. \tag{1}$$

Here N_t represents the number of radioactive atoms present at time t, N_0 is the number of radioactive atoms originally present, e is the base of natural logarithms (about 2.718), and λ is a constant (the *decay constant*) specific to the element whose decay is being considered.

Suppose that, at the time of formation of a rock, P_0 atoms of a parent element and D_0 atoms of one of its daughters were present. Suppose also that the rock neither gives up nor receives additional parent or daughter atoms during the ensuing years and that today P_t atoms of the parent element and D_t atoms of the daughter element are present. Then, by the assumption that parent and daughter atoms neither entered nor exited, we know that the extra daughter atoms that are now present must come from decay of the parent. So we can conclude

$$D_t = D_0 + P_0 - P_t. \tag{2}$$

From equation (1) we get

$$P_t = P_0 e^{-\lambda t}, \tag{3}$$

where t is now the age of the rock. Combining (2) and (3) gives

$$P_t = (P_t + D_t - D_0)e^{-\lambda t}. \tag{4}$$

Elementary algebra will enable us to compute t from this equation, provided that we know P_t, D_t, D_0, and λ. Since constants of radioactive decay are specific to elements, experimental studies of the decay of the parent element in question provide the value of λ. P_t and D_t can both be calculated by measuring the amounts of parent and daughter isotopes found in the present rock. (As Dalrymple points out in a forthcoming work, available techniques give us more than the accuracy we need.) But there is an apparent problem with D_0. How can we figure out the amount of the daughter element originally present? The answer is that in many cases (if we choose the right element for the right rock) we have excellent reasons for believing that D_0 is zero (or, at least, negligibly small). In other cases, as we shall see, we can use present rock compositions to infer the value of D_0.

Obviously, there are two major assumptions involved in the use of radiometric dating. Scientists have to estimate D_0, and they have to rule out the possibility that additional quantities of the daughter element have been added since the time the rock was formed. (Actually, the computation of the age would be affected if some of the daughter element originally present had been lost. However, since we are primarily concerned with the Creationist challenge, the main worry will concern subsequent *additions*. For if extra daughter element were added, then we should arrive at too large a figure for the amount of the parent element that has decayed, and thus produce too high a value for the age of the rock.) Geologists are not unaware of these assumptions, and they take great pains to construct ways of cross-checking them.

Consider first the ways of computing D_0. One common method of radioactive dating, the *potassium-argon method*, takes the radioactive potassium isotope, potassium-40, as the parent element and argon-40 as the daughter element. Argon is an inert gas, so that it does not occur in chemical compounds in original rocks. In some crystalline structures it can be trapped mechanically, but for other naturally occurring minerals it can be shown that this does not occur. Hence, in the case of these minerals, we can conclude that no argon was originally present; that is, $D_0 = 0$. The chief defect of the potassium-argon method is that, under the action of heat or compression, argon can escape from rocks, so that the estimated age is *less* than the true value. A second common method of radiometric dating involves the decay of uranium into lead. Here it is possible to use two decay processes, the decay of uranium-238 into lead-206 and the decay of uranium-235 into lead-207. Furthermore, the amount of lead originally present can be computed by considering another isotope of lead. Lead-204 is present in small quantities in most samples of lead, and this isotope is not itself the product of a radioactive decay process. Hence, by measuring the amount of lead-204 in a rock, geologists can estimate the amount of lead originally present. Given this value of D_0, it is then possible to use either decay process to calculate the age of the rock. If the results agree, they are said to be *concordant*, and geologists are usually confident that concordant ages are the true ages of the rocks under consideration.

The second worry is that extra amounts of the daughter element may enter the system after the original formation of the rock, thus giving the impression that more of the parent element has undergone radioactive decay than has actually been the case. In both the examples I have described, there are ways of checking that such intrusions have not occurred. Minerals can be tested for their capacity to absorb extra argon under experimental conditions designed to resemble their natural environment, and geologists can screen out, in this way, minerals that are liable to give erroneous results. In the second case, the existence of two separate decay processes provides a check on the assumption that the system has not been contaminated. If extra lead were to have been absorbed in the rock after the original formation, the new lead would have caused the calculated ages of the rock to diverge unless it contained the right proportion of lead-206 to lead-207. If the ratio of lead-206 to lead-207 in the newly introduced rock were greater than the ratio of lead-206 to lead-207 found in an uncontaminated system, the method of dating based on the decay of uranium-238 to lead-206 would give a relatively higher value than the method of dating based on the decay of uranium-235 to lead-207. Obviously just the opposite holds when the ratio of lead-206 to lead-207 is too small. Hence someone who supposes that concordant ages are inflated must believe that the contaminating lead contained just the right proportion of the two isotopes.

I want to emphasize that I have only dealt with two of the commonly used radiometric methods, and I have only outlined the most elementary of the checks that geologists use in applying them. (More details can be found in Eicher 1968, chapter 6; and Faul 1966.) From what I have said it might seem that the assignment of ages to rocks is still a bit uncertain. However, I hope that it will help to quell anxieties when I point out that a large number of independent methods have been applied to a vast array of different rocks. The result of this enormous array of tests is a consensus. The ages assigned to various rock strata bearing distinctive types of fossils show extraordinary agreement. The many independent computations of the age of the earth during the last three decades almost invariably yield a figure between 4.2 and 4.8 billion years. Of course, there are occasional puzzling discrepancies. But geologists take these as signs that unanticipated factors have affected the system from which the result was obtained. They know that geological clocks, like other clocks, can go wrong. Frequently, further investigation dissolves the anomaly by showing what the interfering factor has been.

Let us now take up some of the Creationists' attempts to criticize radiometric dating. The main lines of attack are laid down by Morris. He begins by identifying three assumptions of the use of radiometric techniques: "1. The system must have been a closed system. . . . 2. The system must originally have contained none of its daughter component. . . . 3. The process rate must always have been the same" (Morris

1974a, 138). We have already discussed statements *akin* to Morris's first and second assumptions. As will become clear shortly, the status of the third is a little different.

Unsurprisingly, Morris believes that he can provide good reasons for doubting each of these assumptions in the case of *every* application of *every* method. He claims that none of the assumptions is "provable, testable, or even reasonable" (Morris 1974a, 139). Why not? Here are the reasons:

1. *There is no such thing in nature as a closed system.*
The concept of a closed system is an ideal concept, convenient for analysis but non-existent in the real world. The idea of a system remaining closed for millions of years becomes an absurdity.

2. *It is impossible to ever know the initial components of a system formed in prehistoric times.*
Obviously no one was present when such a system was first formed. Since creation is at least a viable possibility, it is clearly possible that some of the "daughter" component may have been initially created along with the "parent" component. Even apart from this possibility, there are numerous other ways by which daughter products could be incorporated into the systems when first formed.

3. *No process rate is unchangeable.*
Every process in nature operates at a rate which is influenced by a number of different factors. If any of these factors change, the process rate changes. Rates are at best only statistical averages, not deterministic constants. (Morris 1974a, 139)

These rejoinders make it apparent that Morris's formulations of the assumptions underlying radiometric dating are only *akin* to the assumptions examined above. When geologists calculate the ages of rocks, they do assume that the system under consideration has remained closed *in one particular respect*. They suppose that none of the daughter element has been added or subtracted. However, this does not commit them to the idea that the system was *completely* closed, that it engaged in no exchange of matter or energy with the environment. Like his memorable argument about the evolving junkyard, Morris's first reply only demonstrates his lack of understanding of basic concepts of physics. The crucial question is whether we can ever be justified in believing that the system was never contaminated by extra amounts of the daughter element. I have tried to explain how geologists can sometimes obtain good evidence for this conclusion.

Similarly, the second point is misguided. Geologists do not have to suppose that the system originally contained none of the daughter element. What is important is that they be able to compute the amount of the daughter element originally present. [Equation (4), let us recall, enables us to compute the age of the rock if we know P_t, D_t, λ, and D_0. Clearly, it is required only that D_0 be *known*, not that it be *zero*. However, for some methods we can calculate D_0 precisely because we can be sure that none of the daughter element could originally have been present; that is, $D_0 = 0$.] Furthermore, Morris is overstating his case when he declares that we can know nothing about the elements originally present. As noted in chapters 2 and 3, it is perfectly possible to have excellent evidence for statements about events and situations that no human has observed. Geologists draw conclusions about the composition of original rocks by applying claims about the possibilities of incorporating elements into minerals, claims that can be tested in the laboratory. So, for example, the thesis that certain minerals would have contained no original argon rests on a perfectly testable (and well-confirmed) claim. While those minerals were in the molten state, prior to the solidification of the rock, argon would have diffused from them. It is only after the molten rock has solidified that the argon formed through radioactive decay becomes trapped within it. Obviously, what is being applied in this case is our knowledge of the physical and chemical interactions of minerals and elements.

Morris's third assumption, and his attempt to undermine it, raises a new issue. In deriving equation (4), from which rock ages can be computed, I employed equation (1), the equation of radioactive decay. I asserted that λ, which measures the rate of decay, is a constant. Morris suggests that the assertion is unwarranted. However, the claim that λ is a constant does not descend out of thin air. It is the result of our knowledge of nuclear physics. Although the sciences *sometimes* teach us that the rate at which a process occurs can be affected by a number of factors, as when we learn that the rate at which water boils is affected by the pressure or that the rate at which mutations occur varies with x-ray irradiation, what we sometimes discover is that a process is impervious to outside influence. Precious little affects the time of passage for a light ray between two points. Similarly, nuclear physics tells us that radioactive decay is well insulated against external interference. The reason is that the emission of particles from an atomic nucleus is under the control of forces that are enormously more effective at short distances than the forces at work in most physicochemical reactions. Moreover, the theoretical predictions that radioactive decay rates are extraordinarily hard to alter are experimentally well confirmed. Extensive attempts to modify these rates under a variety of physicochemical conditions have produced no effects.

When Morris attempts to support his points in the context of particular dating methods, he fares no better. For example, his chief weapon in arguing for the possibility of variable decay rates is a vague proposal that the capture of free neutrons or the impact of neutrinos could affect decay constants (Morris 1974a, 142–143). (The latter idea is linked to a paragraph quoted from a "Scientific Speculation" column.) But neither of these processes would affect *rates* of decay; even granting the possibility of change by neutrino impact or the practical likelihood of neutron capture, the result of these processes would be a modification *not of the decay rate, but of the decaying nucleus.* (The old nucleus, which had been decaying at *its* specific rate, would be changed to a new nucleus, which would then change at *its* specific rate. Note that if processes like these were to occur, they would be detectable since two separate sets of daughter elements would be produced.) Morris's speculations are based on confusion.

Morris then goes on to ignore the methods that geologists employ to ascertain the original amount of daughter element present in the rocks they attempt to date. His discussion of uranium-lead dating contains no mention of the simple technique for computing the initial abundance of lead that I described above. (Needless to say, nothing is said about more sophisticated methods.) His treatment of potassium-argon dating includes the statement: "Since Argon 40 is a gas, it is obvious that it can easily migrate in and out of potassium minerals" (Morris 1974a, 145). However, argon-40 is an *inert* gas, which does not become chemically bound to potassium minerals. Moreover, the crystalline structure of some minerals makes them impermeable to argon. Hence the suggestion that the minerals that geologists date are easily contaminated is simply false.

My brief discussion has only looked at a sample of the objections that Morris and his colleagues (notably Slusher; see Slusher 1973) offer against radiometric dating. The errors I have identified are typical. No attempt is made to criticize the techniques that geologists carefully employ to determine the value of D_0 or to test whether the system has been contaminated. Instead, those techniques are ignored. The picture thus presented is that radiometric dating methods compute the ages of rocks by applying equation (4), assuming dogmatically that D_0 is zero and that the system is uncontaminated. Add to this distortion some vague speculations about changing decay rates (perhaps based on a revisionist nuclear physics under development at the Institute for Creation Research?) and the usual ploy of emphasizing professional disagreements. The result is the typical Creationist mélange—

something that appears authoritative to the inexpert, but can be unmasked by even the briefest account of the standard geological practices.

I shall deal with the positive arguments for a young earth in much less detail. The reason for this is not simply that once one has appreciated the radiometric dating techniques and their overwhelming evidence for the claim that the earth is more than 4 billion years old, it is clear that there must be some flaw in the attempts to show that the earth was created a few thousand years ago. In addition, the Creationist arguments most commonly trotted out share a simple flaw. Creationists assume that certain processes, which we have independent reason to believe to be irregular and sporadic, take place at uniform rates.

Two examples will suffice. Thomas Barnes (1973) argues that the earth's magnetic field is decaying, and he uses the observed rate of decay to compute that prior to about 10,000 B.C. the earth's magnetic field would have been impossibly strong. However, there is overwhelming geophysical evidence for the claim that the earth's magnetic field fluctuates both in intensity and direction, so that Barnes's extrapolation from the present is simply misguided. A similarly erroneous argument is given by Morris (Morris 1974a, 167–169; Morris 1974b 150–154; Wysong 1976, 147–148). He uses the current rate of growth of the human population to calculate the time required for the present population size to be reached from an original pair of individuals. It should be fairly obvious that this is a blunder. There is every reason to believe that the rate of growth of the human population has not been constant, but has fluctuated quite wildly in the past. Indeed, it is surely an oversimplification to consider the growth of *the* human population except during the last few centuries. If we think about the distribution of humans at the dawn of recorded history, then it is far more realistic to conceive the human race as consisting of a number of relatively small populations. Some of these were fairly successful and were able to expand until they reached the maximum size that their local environments could bear. Others were wiped out by disease, dwindling resources, or competition with other groups. The entire human race may be regarded as a single population only for the most recent past; that is, it has only been very recently that humans have had the power, though not necessarily the desire, to redistribute the earth's resources so as to overcome local limits imposed by the local environment.

Barnes and Morris both choose processes that we know to operate at different rates at different times, and then use the observed rates to estimate the time at which the process began. Dating the past is a complicated and technical business, and one cannot ignore the technical details simply to generate the ages one wants. Without a thorough understanding of which rates are constant over time and which rates fluctuate wildly, Creationist dates are bound to be stabs in the dark. However, Creationists know what they want the age of the earth to be. So just as in the case of the second law of thermodynamics, important parts of science are abused. By carefully picking a process on the basis of its ability to give the desired result, without attending to the question whether it is reasonable to think that it happened at a constant rate, Creationists attempt to convince the uninitiated that their blind dates have scientific references. Nobody should be taken in.

Creation "science" is spurious science. To treat it as science we would have to overlook its intolerable vagueness. We would have to abandon large parts of well-established sciences (physics, chemistry, and geology, as well as evolutionary biology, are all candidates for revision). We would have to trade careful technical procedures for blind guesses, unified theories for motley collections of special techniques. Exceptional cases, whose careful pursuit has so often led to important turnings in the history of science, would be dismissed with a wave of the hand. Nor would there be any gains. There is not a single scientific question to which Creationism provides its own detailed problem solution. In short, Creationism could take a place among the sciences only if the substance and methods of contemporary science were mutilated to make room for a scientifically worthless doctrine. What price Creationism?

6

Exploiting Tolerance

A plea for equal time

John Dewey once explained the nature of genuine intellectual tolerance by using a homely analogy. Being open-minded, he suggested, is like placing a welcome mat outside your front door and being prepared to be hospitable to those who ring the doorbell. It is not tantamount to throwing the door wide open and posting a sign "Come on in. Nobody's home." Creationists are not exactly wild about Dewey's ideas. They frequently blame Dewey for converting schools into institutions for "indoctrinating children into a pluralistic, democratic society" (Hefley 1979, 30; Morris 1974b, 33). Yet, as I shall argue, Dewey's simile provides an apt basis for evaluating the Creationist plea for equal time.

Pleas for tolerance are calculated to strike a responsive chord in many people. Falwell's printed letter soliciting (tax-deductible) contributions to aid in the Creation vs. Evolution battle carefully portrays the Creationist movement as advocating open-mindedness: "As you know, we are involved in a massive campaign to inform the American people of the truth—that the concept of special creation should be taught in public schools alongside the concept of evolution . . . in the name of academic freedom" (ellipses in original). Others who fight for changes in curriculum and in textbooks so as to emphasize "traditional values" try, when they can, to depict themselves as the apostles of fairness. For example, Norma Gabler, whose attempts to change textbooks in Texas have probably had more impact than any other grass-roots educational movement of the same kind, presented her appeal for "balanced treatment" in the biology curriculum with the following words:

In appearing before investigating committees we have heard repeatedly that parents have been told that a . . . student must be given both sides of an issue. Now, we as parents, are pleading for the same thing . . . let our children be given both sides in the field of biology

The thing that means the most to me, particularly in light of our recent Supreme Court ruling, [is] our children will be taught atheism out of these books. . . . Why is it unfair to teach the creation theory to a child? Why can't they teach that? (Hefley 1979, 49; ellipses in original)

When the issue is cast in these terms, it is hard for people outside the educational and scientific communities to withhold their sympathy. The power of the plea for tolerance was clearly illustrated in the findings of a recent poll. Of those questioned, 76 percent declared that both "theories"—creationism and evolutionary theory—should be taught in high school. The number of those who favored the teaching of Creationism alone was 10 percent; 8 percent favored the teaching of evolution alone; and 6 percent were undecided. These results should not come as a surprise. When ordinary fair-minded citizens are asked if "both sides of the issue" should be presented, whether "both theories" should be made available, they agree quite readily that the treatment of the topic ought to be unbiased. Few people have sufficient expertise to recognize that the question should never have been asked in those terms, that the alleged "theories" are not genuine rivals.

There are some grounds for doubting that the Creationists are sincere in their desire for "equal time" or "balanced treatment." They campaign for representation of their views in the high school curriculum, asking for as much as they think they can get. However, their writings sometimes suggest that they would like more. The general edition of *Scientific Creationism* declares, "True education in every field should be structured around creationism, not evolutionism" (Morris 1974a, iii). Elsewhere, Morris concludes a section on "The Faith of the Evolutionist" with the following sentence: "Therefore, creationists are saying with increasing clarity these days, creationism should be taught in the public schools on at least an equal basis with evolution" (Morris 1974b, 22). Despite the clear signals that the plea for equal time is only a first step, I shall suppose that Creationists want what they usually say they want. Does a commitment to open-mindedness require us to give "equal time" or "balanced treatment" to Creationism?

The point of Dewey's analogy is to remind us that ideas have to earn the right to our respect. Not every wild speculation deserves our detailed scrutiny. Not every crackbrained proposal merits a place in our curriculum. Certainly, the theory about aliens I cooked up in the last chapter does not. Those observations are commonplace. What is more difficult is to decide whether the sincere beliefs of a sizable group of people (for religious fundamentalism is widespread, even if the belief in "scientific" Creationism is supported by only a small minority) ought to be entertained as a genuine alternative to the views of experts. Once we recognize, with Dewey, that real open-mindedness requires not that we abandon our intellectual standards, but that we use them to examine the credentials of the ideas that others espouse, we should see that the issue cannot be resolved as quickly as Creationist literature frequently suggests. Instead, we must determine what criteria have to be met for a view to secure a place in our educational system and whether Creationism satisfies those criteria. Fortunately, open-mindedness does not (yet) require us to practice doublethink and to adopt the other mind-destroying machinery of *1984*.

The main goal of this chapter will be to present an account of how a rational and tolerant community of citizens, scientists, and educators would proceed, and to apply this account to the case that concerns us. However, some subsidiary issues will also arise. Creationists expend a great deal of energy on attempting to convince the world that they are members of the scientific community and that their debates with evolutionists are internal scientific disputes. These efforts are an important part of the plea for equal time, and, at the end of the chapter, I shall examine this aspect of the Creationist campaign.

Undogmatic science

Creationists are able to exploit a number of popular opinions. In recent years, the public image of the scientific community has become somewhat tarnished. Professional scientists have not done a very good job of popularizing the complicated ideas of contemporary research and making the character of their activity accessible to the general public (Nelkin 1977; there are encouraging signs that some scientists are now taking more seriously the enterprise of explaining the principal ideas of contemporary science). Second, there is considerable popular mistrust of educational policy. Despite my conviction that the Gablers were often wrong in their campaign to reform textbooks, I found it impossible to read an account of their strenuous efforts (Hefley 1979) without coming both to admire their perseverance and to sympathize with their predicament. For, if Hefley's depiction of the situation is even approximately correct, it is clear that the processes of curricular design and textbook choosing, the processes the Gablers tried to affect, were insensitive to the opinions of parents. At least in this case, educational policymakers and textbook publishers were not receptive to criticism from parents, and there is some reason to doubt that their attitude would have been different if the complaints voiced had had a better rationale than those that the Gablers presented. Like most parents, I would hope that my own concerns about the education of my children would not be dismissed with a superior smile.

The third source of the idea that the scientific establishment is closed-minded is a philosophical picture of the nature of science. Thomas Kuhn's book *The Structure of Scientific Revolutions* has probably been more widely read—and more widely misinterpreted—than any other book in the recent philosophy of science. The broad circulation of his views has generated a popular caricature of Kuhn's position. According to this popular caricature, scientists working in a field belong to a club. All club members are required to agree on main points of doctrine. Indeed, the price of admission is several years of graduate education, during which the chief dogmas are inculcated. The views of outsiders are ignored. Now I want to emphasize that this is a hopeless caricature, both of the practice of scientists and of Kuhn's analysis of the practice. Nevertheless, the caricature has become commonly accepted as a faithful representation, thereby lending support to the Creationists' claims that their views are arrogantly disregarded.

Would a truly undogmatic scientific community take Creationism seriously? Setting on one side for the moment the issue of how closed-minded the scientific community actually is, let us inquire how it would proceed if it were genuinely tolerant. I begin from a simple and obvious point. Genuine intellectual tolerance is rooted in a desire to learn the truth. For a person or a group, to be genuinely undogmatic is to recognize the possibility that the doctrines that are presently accepted, even if they are deeply valued and even if they are extremely well confirmed, may for all that be mistaken. Those who wish to arrive at the truth will be on the watch for potential problems with the opinions they espouse. They will be sensitive to the deficencies of their views and will encourage discussion of alternatives that promise to correct those deficiencies. Yet respect for the truth does not require one to take seriously ideas simply because they are popular or backed by influential people. Once it has become clear that a proposal makes no contribution to our understanding, we are not compelled by tolerance to give it further attention.

At any stage in the history of inquiry, we can divide claims into three major categories. In the first are those that are best justified by the evidence we have. These may not be entirely free of problems, but confirmation for them is stronger than for any alternative. So, for example, Einstein's special theory of relativity currently enjoys greater support from the evidence than any alternative theory about the dynamics of bodies. A second category consists of claims that are less well justified than some rivals, but merit our attention because of their ability to resolve some outstanding puzzle. If we had to label these claims as true or false, we would have to reject them. (For they are less well supported by the evidence than the rival view that we accept. Because the views are genuine rivals, incompatible claims, it cannot be that both are true. So we have to reject at least one.) Nevertheless, the heterodox views have some promise. There is some question, or group of questions, that they can tackle more successfully than their orthodox rivals, so that we may reasonably suspect that they may guide us in our search for a better view. Hence these claims, while not worthy to be accepted as true, earn entry into the forum of ideas. The final category comprises the residue. These are claims that, given the available evidence, have nothing going for them. From the standpoint of the present evidence, they are likely neither to be true nor to advance our search for the truth.

Open-mindedness is not incompatible with having opinions. Each person (or group) is constantly required to make decisions, and in decision making we use the views we hold about the way the world works. What is crucial to open-mindedness is the *way* in which opinions are held. An (ideal) open-minded scientific community would carefully weigh the evidence and accept only the best-supported claims. It would continue to discuss those claims that fall into my second category, making sure that any insights they contain are not lost in the enthusiasm for developing the orthodox position. Claims belonging to the third category would be shelved. Moreover, all these evaluations would be subject to revision; the open-minded community would admit that new findings can affect its categorization, perhaps causing it to dust off some claim long on the shelf or to reject a claim that has enjoyed popularity.

In considering the rationality of scientific investigation it is often tempting to look just at the behavior of individuals. Historians and philosophers sometimes attempt to understand the acceptance of a new theory by examining the evidence that led some prominent figure to accept it. (For example, they might ask what evidence Darwin had for accepting the theory of evolution by natural selection.) But to focus only on the attitudes of individuals and the evidence for those attitudes is to miss important questions about scientific inquiry. An individual must concentrate on one line of research, the avenue that appears most promising in the light of the available evidence. But the community can be more flexible. It can encourage pursuit of a number of alternatives, expending most of its resources on articulating the position that is best supported, but allowing views with some promise to develop themselves.

A group of scientists deeply committed to discovering the truth would manage its affairs by a division of labor: Most members of the group would be encouraged to work further on the theory that is best justified by the evidence. But at the same time some would be encouraged to actualize whatever promise alternative theories may have, and others periodically to examine the categorization of theories in light of changing evidence. (Perhaps it is also important that the history of science be studied, not only because "dead" theories are sometimes resurrected but also because history is a powerful reminder that the mightiest theories may fall.) If there should come a time when the evidence is inconclusive between two rival theories, then roughly equal numbers of (equally talented) scientists should be encouraged to work on each of the rivals. A community that conducted itself in this fashion (perhaps by creating professional incentives to promote an optimal division of labor) would be, I suggest, both rational and undogmatic. (Note that the individual members of such a community may be quite single-minded in their pursuit of the particular line of research "assigned"to them.)

I have tried to give a sketch of what undogmatic science would be like. We can now take up two questions. Is the actual practice of the natural sciences at all similar to the imagined behavior of my ideally rational and tolerant community? Would an ideally rational and tolerant community, given the evidence available today, commit any of its resources to discussing and developing the Creationist program?

There is some indication that, at least at some times, groups of scientists approximate the ideal just described. In 1915 Alfred Wegener, a meteorologist, suggested a new geological theory, the theory of continental drift. According to this theory, the continents have not always occupied their present positions. Rather, they were once united to form a single landmass, which was broken up at some point in the distant past. Wegener used these ideas to account for some puzzling facts about the shapes of the continents, the apparent continuity of mountain ranges on different continents, the distribution of plants and animals, past climatic conditions, and so forth. His careful attention to detail made it clear that the new theory could answer some questions that were unanswered by rival theories. However, Wegener's proposal faced an apparently insuperable difficulty. How do the continents move? Wegener's own attempts to suggest a mechanism for continental motion were plainly inadequate, and the difficulty was compounded by the well-known fact that continental rock is less dense than the rock at the bottoms of oceans. Geologists concluded that, for all its promise, Wegener's theory could not be right. The "light" continents simply could not plough their way through the "heavy" oceans.

About half a century later, the new theory of plate tectonics proposed that the continents rest on huge plates that can move apart, come together, or slide past one another. In this way, the theory equipped Wegener's proposal with the needed mechanism. It also showed how large numbers of puzzling facts could be explained if the new theory was accepted and won almost universal support for continental drift. What interests me here is not the way in which the old, discredited idea suddenly won allegiance, but how it survived. Despite the fact that continental drift had seemed impossible, Wegener's proposal did not simply disappear from view. Periodically, there were scientific discussions of its merits. Some geologists—including at least two respected scientists (Holmes and du Toit)—worked on developing it and published papers in its defense. Young geologists heard about the idea in geology courses. In short, professional discussion of a minority view was allowed. At times it was even fostered. Moreover, when the balance of evidence began to shift, the geological community was quick to abandon its old orthodoxy and to accept Wegener's apparently wild idea. (For a lucid account of the episode, see Hallam 1973.)

An ideally rational and open-minded scientific community, faced with Wegener's theory and Wegener's evidence, would have done what the actual scientific community actually did. Because of its ability to address some unresolved problems, Wegener's proposal earned the right to scientific discussion. Because Wegener was clear and definite in stating his theory, his views *could be* discussed. Because this theory encountered a grave difficulty, it did not displace the orthodoxy of the 1920s, but it persisted as a minority view.

The contrast with Creationism is striking. Creationism does not merit scientific discussion. As we found in the last chapter, Creation "science" is not a promising rival to evolutionary theory. It is not integrated with the rest of science, but is a hodgepodge of doctrines, lacking independent support. It offers no startling predictions, no advance in knowledge. We cannot commend it for any ability to shed light on questions that orthodox theories are unable to answer. Nor can we praise it for offering a definite alternative that might help scientists in their quest for an improved biological or geological theory. "Scientific" Creationism has no evidence that speaks in its favor, partly because Creationists are so meticulous in leaving their doctrines blurred.

Finally, there is no excusing it on the grounds that its resources are, as yet, untapped. Ample opportunity has been provided. Numerous talented scientists of the eighteenth and early nineteenth centuries tried Creationism. Nothing has come of their efforts, or the efforts of their modern successors. Where the appeal to evidence fails so completely, the appeal to tolerance cannot succeed.

Open-minded education

So far I have discussed the proper attitude toward Creationism of an undogmatic scientific community. However, the main issue is not what consideration should be given to the doctrine by professional scientists — the National Science Foundation is not about (voluntarily) to divert resources toward the Creationist program — but what high school science teachers ought to teach. I shall use my conclusions about undogmatic science in considering whether open-mindedness requires us to include Creationism in the science curriculum.

The aims of high school education in the sciences are not exhausted by schooling future professional scientists in the rudiments of various scientific fields. Students who go on to any number of different occupations will benefit from some basic scientific education. They will vote in a country where important issues are often technical. If they are to understand those issues, then they must have some elementary knowledge of the natural sciences, some familiarity with the vocabulary of science, some capacity for seeing how they can obtain information that they do not have. There are also less practical benefits. Natural science has advanced our understanding and appreciation of so many different aspects of the universe that it would be outrageous if our educational system failed to convey even an inkling of what has been accomplished. Moreover, the study of science provides important training in reasoning. Even a brief look at the ingenious experiments and arguments on which scientific claims rest can help students to think more clearly.

I shall not try to derive my claims about the purposes of science education from some grand general theory of education. I think that they carry more conviction on their own merits, as relatively obvious points about what high school education in science attempts to do. Let us now ask how each of the four functions that I have identified would be affected by introducing Creationism into the high school science curriculum.

The education of professional scientists would clearly be affected adversely by taking time from teaching basic science in order to consider a doctrine that a rational and ideally tolerant scientific community would find worthless. If "scientific" Creationism merits no discussion in the community of professionals, then it does not deserve a place in the classrooms where those professionals are being educated. This is not to deny that professional education in the sciences might not benefit if it were more open to heterodoxy, if received opinion were not sometimes subjected to pressure from minority views. But the ideas in question ought to have something in their favor. They should not fail so abjectly as Creation "science" does.

A similar point holds when we consider the education of those who, as future citizens, will be asked to vote on technical issues. We have a duty to them and to their fellow citizens to provide them with the best information we can. If students were taught that the atom cannot be split, they would have some difficulty in understanding the issues raised by nuclear technology. If they were taught that intelligence is correlated with race, they might well be misled into callous social views. Knowledge of science can have great impact on social and political policy. Students need to be told, clearly and directly, what statements are supported by the available evidence. It is not the teacher's function to offer instead a contrived and *unresolved* "debate" in which one of the parties is an ill-defined position that lacks any evidence in its favor. To represent as equal ideas of unequal merit is to mislead and confuse. Because the consequence of so deceiving the students may be their later inability to perform their duties as conscientious and informed citizens, such educational practices ought to be recognized for the irresponsible charades they are.

We also mislead and confuse when we pretend that evolutionary theory does not rank with the great achievements of natural science. To suggest that Creationism is a viable alternative to that theory not only interferes with the recognition of important truths about the organic world. It also depicts as a genuine science a doctrine that only feigns science. If we aim to show students the major parts of our knowledge of nature, then evolutionary theory must be included. Nor should we conceal its status by masquerading a piece of pseudoscience as a "scientific alternative" to the evolutionary account.

Finally, giving "balanced treatment" of Creation "science" would subvert the valuable function of teaching students the methods of scientific reasoning. If the fallacies and pretentions of "scientific" Creationism are presented, *and left unanswered*, then we are practicing doublethink. There should be no objection, however, to a certain use of Creationist arguments, one which will not appeal to Creationists. Important purposes may be served by exposing the differences between science and pseudoscience, by showing, as I have tried to show in preceding chapters, how Creation "science" consistently falls short of the standards required of the genuine article. Trouble only threatens when serious lines of reasoning (the careful arguments for evolutionary theory and the scientifically responsible challenges to neo-Darwinian orthodoxy) are left to dangle with the fallacies that the Creationist offers—and when the teacher exits without explaining the difference.

My conclusion can be summarized in a sentence. It is educationally irresponsible to pretend that an idea that is scientifically worthless deserves scientific discussion.

Aggressive Creationists sometimes ask why educators are afraid to let the Creationist case be heard. They contend that students ought to be allowed to resolve the "Question of Origins" for themselves. Falwell's letter puts the point this way: "Too many of our children have heard only the evolution side of the story. And . . . evolution is being taught as absolute fact . . . not as theory. This is crucial because knowing the truth about creation definitely affects how you live on this earth, and how you relate to your fellow man" (ellipses in original). For the moment, let us grant Falwell his evaluation of the significance of the issue. (The question whether evolutionary theory has an impact on human conduct will be considered in the next chapter.) Let us also concede that *if* teachers do not point out that science, as a whole, is a fallible enterprise, that all our major achievements, from cosmology, chemistry, evolutionary biology, or whatever area of science, might need revision, *then* they have failed to convey an important characteristic of natural science. What is wrong with Falwell's reasoning is its pretense that there is an issue here with two sides, both of which ought to be presented in the classroom.

Yet, one may ask, why not let Creationists submit their case? Surely, the truth will out, and, if evolutionary theory is indeed well supported by the evidence, then the scientific and educational establishment should have nothing to fear. Nor should we worry about a little wasted classroom time, when a deeper understanding of the merits of evolutionary theory might be secured by allowing students to think through the issue for themselves. The argument is insidious, and Creationists delight in making it.

Even now it is not too late to restore to our children their rightful heritage—the whole Bible as the only sure basis for education in religion, in history, *and in science.*

Or, at the very least, the children should be informed that there are *two conflicting views on origins: some* scientists believe in evolution, others in special creation. The arguments on *both* sides should be presented, and children should be *free to choose* between them.

The present dogmatic teaching of evolution guesswork as "fact" is closely akin to brainwashing, and is indefensible on any principle of logic, ethics, or democracy. (Watson 1976, 103)

Let's begin giving students a choice by making "evolution" and "creation" optional subjects, rather than continuing to cram evolution down every student's throat in every subject as the only posible explanation for life. . . . Could it be the evolutionists are afraid that their case cannot stand comparison [with creation]? If they had confidence in their stand, they should gladly welcome all available evidence. If their case is this weak, why allow it to be forced on students? This isn't fair to the thousands of students who believe in creation. (Mel and Norma Gabler, quoted in Hefley 1979, 147)

It is especially important in these subjects [history, social sciences] that the teacher gives a balanced presentation of both points of view to students. Otherwise, the process of education for living becomes a process of indoctrination and channelization, and the school degenerates into a hatchery of parrots. (Morris 1974a, 178)

In all three cases, we encounter the charge that, unless both "models of origins" are fairly presented, education becomes indoctrination. The Gablers make the further suggestion that a decision not to present the case for Creationism would stem from a cowardly refusal to face the truth. (On this point, see also Wysong 1979, 27–29.)

Evolutionary theorists and educators do not fear the evidence. There is no doubt that a fair presentation of the evidence, and a careful review of the arguments, will support evolutionary theory and unmask "scientific" Creationism for what it is. The previous chapters show that there is no genuine contest, no true comparison. What is in doubt is the possibility of a fair and complete presentation of the issues discussed above, in the context of the high school classroom. Morris advertises *Scientific Creationism* as a work that can "equip the teacher to treat all of the more pertinent aspects of the subject of origins" (Morris 1974a, 3). So we can expect that the discussion of Creationism envisaged will consist of a rehearsal of the arguments whose foibles have been examined. There will be little emphasis on the positive achievements of Creationism (for there are none) and much dredging up of misguided objections to evolutionary theory. The objections are spurious—but how is the teacher to reveal their errors to students who are at the beginning of their scientific studies? (Would it even be permissible for the teacher to expose Creationist distortions?) As we have seen, Creationists scatter their criticisms, using whatever ammunition they can find. Even a gifted teacher would not be able to expound enough of the scientific background to make it clear that all the salvos miss the mark.

What Creationists really propose is a situation in which people without scientific training—fourteen-year-old students, for example— are asked to decide a complex issue on partial evidence. Creationists can purvey their rhetorical wares for a sufficiently long time—recall the abuse of thermodynamics, the misreadings of the fossil record, the vague appeals to the Flood, the obfuscating calculations about the time needed for evolution, the distortion of methods of radioactive dating. They can make enough criticisms to prevent a biology teacher from identifying all the errors. In short, they can muddy clear waters.

Students would be indoctrinated if they were offered a single view as authoritative when rival views were equally well confirmed by the available evidence. Nothing like indoctrination occurs when the best-supported account of the origin and development of life is presented for what it is. As we have seen, evolutionary theory receives over-whelming support from a diverse body of evidence. It explains the characteristics of organisms, the relationships among groups of organisms, the distribution of plants and animals, the features of the fossil record. "Scientific" Creationism does none of this. It is an indefinite doctrine that makes up for its paucity of problem-solving success by hurling misleading objections at its intended rival. No parent has the right to ask a teacher to disguise these facts, to mouth specious criticisms and to feign belief that they are serious arguments, to sit idly by while students "decide" an issue that they are in no position to resolve rationally.

The appeal to tolerance sounds very persuasive. Students are simply to be offered "both sides of the story." Yet there is a powerful antidote. If the Creationists obtain what they say they want, then responsible teachers will be required to deceive their students. It is small wonder that, in his testimony at the Arkansas trial, a sincere and unassuming science teacher declared his refusal to act in accordance with the law. "I am not a rebel," he said, "but I cannot do it." Teachers have the right, as well as the responsiblity, to give their students as clear a picture of the way the world works as we now possess.

Although Creationists usually argue that Creationism ought to be given a place in the high school science curriculum, their basic demand is really "Both or neither." Hence, we should consider whether the goals of open-minded education might not better be served by deleting from the curriculum any mention of evolutionary theory. Interestingly, in a discussion of the constitutional issues, Morris offers an argument that seems to favor this suggestion, although he draws a different moral:

Since creationism can be discussed effectively as a scientific model, and since evolutionism is fundamentally a religious philosophy rather than a science, it is clearly unsound educational practice and even unconstitutional for evolution to be taught and promoted in the public schools to the exclusion or detriment of special creation. (Morris 1975, 14)

These constitutional provisions [the First and Fourteenth Amendments] clearly preclude the imposition of exclusively evolutionary teachings on the children of parents whose religious convictions favor creationism, since this in effect amounts to compulsory indoctrination in a state-endorsed religion. Creationist children and parents are thereby denied "equal protection of its laws" and the State has, to all intents and purposes, made a law establishing the religion of evolutionary humanism in its schools. (Morris 1975, 14)

Apparently, Morris is either not serious about his claim that evolution is religion or he is not serious about the constitutional amendments he invokes. For he concludes that both "models of origins" should be taught (Morris 1975, 16). Yet, if the Constitution prohibits the teaching of religion in public schools and evolution is a religion, then the correct conclusion is that evolution should not be taught, rather than that *one* cherished alternative should be taught as well. Further, as many people have observed, there is only one consistent way to apply Morris's principle. Once the door has graciously been opened to one particular version of Christianity, other religions must also be allowed to enter. If recognizing that "evolution is a religion" compels us to introduce Creationism into the high school science classroom, then the "sciences" of Moslems, Hindus, Buddhists, and even Druids, cannot be debarred.

However, we are not swept into so absurd a conclusion. For one thing, as seen in chapter 2, the charge that evolutionary theory is a religion will not stick. Moreover, as will be seen in the next chapter, evolutionary theory does not have the broad implications for religious doctrine that Creationists try to attribute to it. Nevertheless, the idea of removing evolution from the high school curriculum may seem to be a simple way of resolving the political debate about the teaching of evolution. That idea has sometimes been popular. In the wake of the Scopes trial, high school biology textbooks were reorganized so as to remove references to evolution or to concentrate them in a single chapter. That policy remained in effect until the 1950s, when the challenge of Sputnik rejuvenated American high school science education. Even though the Creationists have no case, the vehemence of their protests and the strength of their political alliances may induce publishers to retreat and to produce books that place little or no emphasis on evolution.

It would be an intellectual disgrace for that to occur. Evolutionary theory is the heart of biology, a part of the subject that informs all other biological studies. No serious treatment of any group of organisms can avoid mention of their evolutionary history. For it is that history that makes comprehensible their present characteristics and relationships. But, most important, to remain silent on the issue of evolution would deprive students of their right to learn about one of the greatest intellectual achievements of the human race.

Just as one should not pretend that Creationism and evolutionary theory are genuine rivals, so too one should not omit reference to the subject as if it were irrelevant or controversial. Moreover, bowdlerized *biology* would not be the end of the matter. For if the Creationists are taken seriously enough to warrant subtraction of evolutionary ideas, then other scientific texts would have to fall in line. The geological timescale, of course, would have to go, as would those parts of astronomy that suggest that the universe is very old. Chemistry books would have to tread warily around the issue of the formation of biological molecules, and physics texts would not dare to insist that radioactive decay rates are constant. We should be clear about what exactly we would be doing. Removing evolutionary theory from the textbooks and the classroom would be an act of political censorship. Educators would be withholding an important scienfitic theory, with all the evidence on its side, simply because of political pressure exerted by opponents of that theory.

Credential mongering

In presenting the case for teaching their favored doctrine, Creationists adopt certain tactics that are designed to make them appear as dissenting scientists. The rest of this chapter will consider three popular devices. I shall start with an obvious feature of Creationism, the emphasis on scientific credentials.

Virtually every Creationist work that I have read loudly proclaims its author's qualifications and insists that he is not a rarity among Creationists. "DR. HENRY MORRIS is recognized as one of America's greatest authorities on scientific Creationism. He is thoroughly equipped to come to grips with his subject material. Armed with three earned degrees (including a Ph.D.) in the sciences he served as department head or professor at four famous institutions, Louisiana University, the University of Minnesota, Rice University, and Virginia Polytechnic Institute" (Morris 1972, back cover). "There are hundreds, perhaps thousands, of scientists today who once were evolutionists but have become creationists in recent years" (Morris 1972, iv). "Dr. Wysong is a graduate of Michigan State University where he received both his B.A. and D.V.M. He is a practitioner of veterinary surgery and medicine and has also taught part time" (Wysong 1976, back cover). "Dr. Duane T. Gish is a careful scientist of impeccable academic credentials" (Morris's foreword to Gish 1979). And so it goes, every dropping of a "Ph.D." designed to suggest that there is a substantial number of trained scientists who defend Creationism. Creation-Life Publishers even publishes a booklet entitled *21 Scientists Who Believe in Creation*.

As with the "scientific" arguments, Creationist claims about credentials look better when presented in soft focus. Morris's claim about the sheer number of Creationist "scientists" is a wonderful rhetorical ploy—*perhaps* there are thousands. Then, again, perhaps not. More important, while the Creationists whose credentials are flaunted on fliers and dust jackets do have various degrees, by and large these degrees are not in the *relevant* fields. On closer inspection, the "21 scientists who believe in Creation" hardly constitute a distinguished panel of experts on the origins of life: three hold doctorates in education; two are theologians; five are engineers; there is one physicist, one chemist, a hydrologist (Morris), one entomologist, one psycholinguist, and someone who holds a doctorate in Food Science Technology; finally, there are two biochemists (including Gish), an ecologist, a physiologist, and a geophysicist. While the last five may have some expertise in related areas, the credentials of the others are utterly irrelevant to many of the questions Creationists address. The "authority" of these men should not convince us that there is a scientifically reputable alternative to a major *biological* theory. The word of just any " scientist" is not enough. I am prepared to bet that Creationists, like the rest of us, take care to consult the *appropriate* experts. I doubt that they take their sick children to the vet.

In any case, the crucial issue is not whether some people who possess doctoral degrees *say* that there is a case for Creationism, but whether they are *right* in saying so. To settle that issue, it is wiser to look at the evidence itself. We have examined the evidence, and found it sadly deficient. That detailed evaluation should not be disturbed when we learn that some men who have achieved doctorates in various fields of science think otherwise. Mere possession of a Ph.D., even in the relevant field, does not guarantee that a person's views are correct, or even defensible. Scientists must continually earn the respect of the scientific community by the quality of their work.

Genesis without God

The second tactic used by Creationists is to present their "model of origins" as if it had no commitment to any religious doctrine. Quite understandably, Creationists are sensitive to the charge that what they are really demanding is time for instruction in a particular religion. The "public school" edition of *Scientific Creationism* carefully avoids any references to the Bible in its seven chapters. The "general" edition of the same work adds an eighth chapter in which various religious points are pursued. This relation between the two works raises an interesting question: Should we read the final chapter of the "general" edition as a religious appendix to a scientific work? Or is the "scientific" treatment just a specially contrived version of a religious tract?

Creationists, of course, want to insist that the seven chapters of the "public school" edition stand on their own feet. Since it is difficult to undermine this claim, hard to show that the religious tail really wags the "scientific" dog, I have taken them at their word. Hence, my previous discussions focus on the scientific merits of the arguments against evolution and the evidence for Creationism. Nevertheless, the question remains. Is there any evidence that might lead Creationists to amend their "scientific" claims, so that they no longer consist simply of a censored version of a literal reading of the Genesis account? (Recall the gratuitous assumption that, after the Flood, all land animals radiated from Mt. Ararat.) If there is not, then Creation "science" is, at bottom, a religious doctrine. In the pursuit of real science, no part of current theory is beyond question. Anything is potentially revisable. To demand that certain parts of "scientific" Creationism must be kept fixed, whatever the evidence, is to drop even the pretense of doing science.

Since contemporary Creationists do not provide detailed answers to questions, since they do not take seriously the potential scientific objections, there is no definite indication of how they might revise their "science." Yet there are moments when the guard slips, as, for example, when Morris announces that "Bible believing Christians" are "forced" to believe that the standard measurements of the age of the earth "have somehow been misinterpreted" (Morris 1972, 89). Watson is even more candid: "Just as Christ at Cana compressed into one second the process of thirty years, so in one day He flung the stars billions of light-years into space, at the same time causing their light to fall upon the earth. How do we know? Because He says so!" (Watson 1976, 31). Thus the puzzling astronomical fact that some observed stars are so far away that their light would take more time to reach us than is available on the Creationist account is dismissed. A special act of creation (the creation of light that *appears* to come from those stars) is invoked, on the grounds that the Bible tells us that something of the kind must have occurred. Finally, Gish neatly sidesteps any awkward questions about how the Creator operated: "We do not know how God created, what processes He used, *for God used processes which are not now operating anywhere in the natural universe*. This is why we refer to divine creation as special creation. We cannot discover by scientific investigations anything about the creative processes used by God" (Gish 1979, 42).

All of these passages provide *evidence* for a judgment: Creationists will not allow any observational findings, any problematic result, to modify the fundamentals of their account; the concordance of Creation "science" with the Genesis narrative (literally interpreted) must always be preserved. The evidence is not conclusive. (Of course, it never is.) Despite all their willingness to invoke miracles and supernatural processes, *perhaps* there is some observational result that might lead Creationists to believe that their account must diverge from their reading of the Bible. But if there is not, then "scientific" Creationism is simply Genesis without God (or, more precisely, a fundamentalist reading of Genesis without God), a theologically disinfected version of a religious doctrine that has no claims to count as science.

Misleading quotation as a way of life

The final tactic that I want to consider is one that we have encountered from time to time in reviewing Creationist arguments. Because Creationists would like to identify themselves as members of the scientific community, scientists engaged in an internal debate with other scientists, they pounce on any remarks by eminent biologists or geologists that can be made to suggest their point of view. These remarks are wrenched out of context—whether Creationists simply do not realize the importance of the context or whether they are willfully distorting the author's intentions, I do not know. In any case, for the Creationists, misleading quotation has become a way of life.

During the Arkansas trial, Stephen Gould pointed out one example of misleading quotation. In discussing the Lewis overthrust in Montana, Whitcomb and Morris want to suggest that the underlying shales have been undisturbed, so that geologists might legitimately conclude that older rock has *not* been forced over younger rock. They quote from an 1886 report, in which the following sentence occurs: "Most visitors, especially those who stay on the roads, get the impression that the Belt strata are undisturbed and lie almost as flat today as they did when deposited in the sea which vanished so many years ago" (Whitcomb and Morris 1961, 187: footnote 1). As Gould remarked, the reference to "staying on the road" suggests that a contrastive sentence might be coming. And indeed, the next sentence explains that those who take a closer look can see evidence of disturbance. Whitcomb and Morris do not quote that sentence. Perhaps this is because it conflicts with the point they are trying to defend. Or perhaps we should accept the explanation that Gish is reported to have offered at the Arkansas trial: "After all, you have to stop quoting somewhere."

Another place in which Creationists use the device of misleading quotation is in discussions of human evolution. Students of fossil hominids are almost all agreed that the *australopithecines* (hominids who lived about 3 million years ago) walked upright. One important source of evidence is the structure of the pelvis, which is extremely similar in *Australopithecus* and *Homo sapiens*. (A summary of the evidence is available in chapter 5 of Campbell 1974; see also le Gros Clark 1964 and le Gros Clark 1971.) Morris ignores this vast body of evidence, neglects to mention that there is a consensus among anthropologists that *Australopithecus* had a bipedal gait, and quotes a single paragraph from volume 100 of *Science News* (1971, 357): "*Australopithecus* limb bone fossils have been rare finds but Leakey now has a large sample. They portray *Australopithecus* as long-armed and short-legged. He was probably a knuckle-walker, not an erect walker, as many archaeologists presently believe" (Morris 1974a, 173). On the basis of a single dissenting judgment, the conjecture of a single unidentified author, Morris feels entitled to bypass the work of anthropologists over half a century, and to conclude that *Australopithecus* "walked like an ape" (Morris 1974a, 173).

Dealing with the same issue, Gish refers to papers by Charles Oxnard in which Oxnard argues that the relations between *Australopithecus*, *Homo*, and the apes are quite complicated. Gish quotes the first sentence of the summary of an important review article that Oxnard published in *Nature*: "Although most studies emphasize the similarity of the australopithecines to modern man, and suggest, therefore, that these creatures were bipedal tool-makers at least one form of which (*Australopithecus africanus*—"*Homo habilis*," "*Homo africanus*") was almost directly ancestral to man, a series of multivariate statistical studies of various postcranial fragments suggests other conclusions" (Oxnard

1975, 389). Gish does not quote the remaining sentences of the summary: "Their locomotion may not have been like that of modern man, and may, though including a form or forms of bipedality, have been different enough to have allowed marked abilities for climbing. Bipedality may have arisen more than once, the Australopithecinae displaying one or more experiments in bipedality that failed. The genus *Homo* may, in fact, be so ancient as to parallel entirely the genus *Australopithecus*, thus denying the latter a direct place in the human lineage" (Oxnard 1975, 389).

As Oxnard's summary indicates, and the detailed argument of the article explains, the evidence is not for Gish's conclusion that australopithecines were "aberrant apes" (Gish 1979, 123), but for a more complicated picture of the relationships among the hominids than is usually presented. Oxnard does *not* conclude that *Australopithecus* walked like an ape. Even though he stresses some similarities with the orangutan, he offers the following statement: "And because similarities with the orang-utan are only in some anatomical regions and not in others, because the overall composition is mosaic in nature, it is clear that the actual overall mode of locomotion of the orang-utan today is not the model for these creatures" (Oxnard 1975, 394). Gish's gloss on the article presents the conclusion he would like Oxnard to have drawn: "From his results Oxnard concluded that *Australopithecus* did not walk upright in human manner but probably had a mode of locomotion similar to that of the orang" (Gish 1979, 122). The phrase "did not walk upright in human manner" is a masterpiece of deception. True enough, Oxnard denies that the bipedalism of australopithecines is the same as that of humans. But he does not argue that australopithecines did not walk upright (that is, in the—gross—manner of humans). Of course, the statement about the orang is less artistic deception. Here Gish simply contradicts Oxnard's careful words.

One of the Creationists' favorite passages in the paleontological literature is a discussion of the origin of higher taxonomic categories by G. G. Simpson. Here are Simpson's words: "In spite of these examples, it remains true, as every paleontologist knows, that *most* new species, genera and families, and nearly all categories above the level of families, appear in the record suddenly and are not led up to by known, gradual, completely continuous transitional sequences" (Simpson 1953, 360; cited by Morris 1974a, 79–80). Morris quotes this passage to support his view that scientists have "no evidence that there have ever been transitional forms between . . . basic kinds" (Morris 1974a, 79). But Simpson's reference to previous examples ought to give us pause. The paragraph before the quoted sentence contains the following remarks: "Among the examples are many in which, beyond the slightest doubt, a species or genus has been gradually transformed into another. Such gradual transformation is also fairly well exemplified for subfamilies and occasionally for families, as the groups are commonly ranked" (Simpson 1953, 360). Moreover, Simpson goes on to describe some of these examples and to give a summary of the character of the fossil record for mammalian orders (Simpson 1953, 360–376, summary on 368–369). His arguments are directly contrary to Morris's claim, but, instead of giving the sense of the discussion, Morris uses an isolated sentence to suggest that a great paleontologist supports the Creationist conclusion about the fossil record.

As a final illustration, let us look at Gish's most recent treatment of the reptile-mammal transition. Gish begins with a misleading comment about a different case, the reptile-bird transition. After noting that evolutionary theorists usually take *Archaeopteryx* to be a transitional form between reptiles and birds, he writes, "Gould and Eldredge exclude *Archaeopteryx* as a transitional form, calling it a strange mosaic which doesn't count as a transitional form . . . " (Gish 1981, ii). At this point there is a reference to a paper by Gould and Eldredge concerned with elaborating the theory of punctuated equilibrium (Gould and Eldredge 1977). Here is what Gould and Eldredge actually said: "At the higher level of evolutionary transition between basic morphological designs, gradualism has always been in trouble, though it remains the 'official' position of most Western evolutionists. Smooth intermediates between *Baupläne* are almost impossible to construct, even in thought experiments; there is certainly no evidence for them in the fossil record (curious mosaics like *Archaeopteryx* do not count)" (Gould and Eldredge 1977, 147). Note that Gould and Eldredge do not deny that *Archaeopteryx* is a transitional form between reptiles and birds, in the sense that it is an evolutionary descendant of reptiles with some avian characteristics. What they are concerned to oppose is the existence of forms showing a smooth change of design between two distinct morphological patterns. *Archaeopteryx*, they maintain, is a *mosaic*, not a smooth and uniform blend of the features of birds and reptiles. In their view, the patterns exemplified by organisms change abruptly, although many elements in the new patterns testify to the evolution of the new organisms from the old form.

Gish's entire pamphlet consists of a broad application of the same tactic. He presents parts of the standard account of mammalian evolution (the transformation of the jaw joint, reviewed in chapter 4), drawn from the recent work of Crompton and his associates. This account is then confronted with scattered quotations from *earlier* sources (specifically, articles by Kermack and his colleagues) that are portrayed as the latest word on the subject: "Kermack and his co-workers now reject this idea . . . " (Gish 1981, v). However, the articles by Kermack to which Gish refers were published in 1968 and 1973, *in advance* of the very detailed work by Crompton, Jenkins, and others on the evolution of the mammalian jaw and middle ear (Crompton and Jenkins 1979; Crompton and Parker 1978). Moreover, although he cites the works of these authors, Gish makes no attempt to represent their content. Instead he asks *rhetorically* questions about jaw musculature to which Crompton and others have given detailed answers.

So it goes. One scientist after another receives the Creationist treatment. Any qualifying comment, any deviation from orthodoxy is a potential target. Ripped from its context, it can be made to serve the Creationists' purpose, namely, to convince the uninitiated that Creationist theses are sometimes advanced by scientists in scientific debates. But anybody can play the same game. In conclusion, I cannot resist turning the weapon against the Creationist who has used it to greatest effect. Referring to the controversy about transitional forms, Gish writes, "There should be no room for question, no possibility of doubt, no opportunity for debate, no rationale whatsoever for the existence of the Institute for Creation Research" (Gish 1981, ii). How true.

7

The Bully Pulpit

A forced choice

Evolution is the root of atheism, of communism, nazism, behaviorism, racism, economic imperialism, militarism, libertinism, anarchism, and all manner of anti-Christian systems of belief and practice.

(Morris 1972, 75)

We come, at last, to the real problem. Creationists are not down on Darwin because the methodology of evolutionary biology offends their scientific sensibilities. They are not clamoring for reform in biology because they think that important theoretical and empirical breakthroughs await us if we adopt their preferred perspective. The root of the trouble is that the theory of evolution contradicts a literal reading of the first eleven chapters of Genesis. However, contemporary Creationists do not present the issue in quite this way. Their basic strategy is to portray evolutionary theory as opposed to a vast array of valued institutions: family, morality, religion, even science itself. The rationale for this strategy is obvious. By depicting evolution as opposed, not just to a particular doctrine of a particular sect, but to a large number of institutions that are cherished by a large number of people, they make it "easier" for their audience to choose sides in the genuine conflict between evolution and Creationist fundamentalism.

Although this strategy is unavailable in the official debate about what to teach in high school courses, there is no point in pretending that it lacks influence. When all the diatribes about thermodynamics, falsifiability, and the fossil record have been forgotten, the suggestion that evolutionary thinking leads to degeneracy and the dissolution of

This chapter was written jointly with Patricia Kitcher.

society may capture the heart. Those who have been beguiled into thinking that a high school course in evolutionary biology is the gateway to a life of violence and depravity are not likely to ponder the scientific credentials of the theory of evolution.

There are numerous passages in Creationist literature that portray evolutionary theory as opposed to morality and religion and that emphasize the evil consequences of teaching evolution. Morris elaborates the theme at great length, holding "evolutionary philosophy" responsible for promiscuity, pornography, and perversion (Morris 1974b, 166–168). In similar vein, Watson claims that the "new cosmology" lays a firm foundation for "the new theology and new morality" (Watson 1976, 37), and Hiebert asserts that there is an "irreconcilable difference between the Bible and evolutionary dogma" (Hiebert 1979, 145, and see also 17–18). The Gablers complain that textbooks that teach evolution "undermine the faith of thousands of students"; one of their allies, the newspaper editor Reuel Lemmons, suggests that adoption of books presenting evolutionary theory would "entrench sheer atheism" in the classroom (citations from Hefley 1974, 44). Finally, John Moore draws up a table to show how the evil tentacles of evolutionary theory have crept into all corners of modern thought; the sins of Marx, Keynes, Freud, Dewey, Frankfurter, Nietzsche, Camus, and Sartre (among others) are all to be laid at Darwin's door (Moore 1974, 27).

Anxiety about evolutionary theory is coeval with the theory itself. Shortly after the *Origin* was published, the wife of an Anglican bishop expressed the hope that, even if the theory of evolution was true, it would not become widely known. Such muddled nervousness persists, and the Creationists are adept at exploiting it. It would be impossible to consider all the accusations that evolutionary theory fosters evil. However, we shall consider the two most central attacks: the claim that one cannot accept Darwin's theory and be a good Christian and the claim that the theory of evolution purveys an "animalistic amorality."

We should emphasize that the charges about to be examined are not part of the "scientific" defenses of Creationism. The bullying from the pulpit is reserved for the pulpit (whether found in a church or a television studio), campaign fliers, and the "general" (nonscientific) works written by Creationists. The "public school" edition of *Scientific Creationism* contains only the "scientific" arguments lately considered. But other books mix such arguments with religious exhortations. The short popularization of Creationism, distributed by the Old Time Gospel Hour, is quite explicit on the point that the scientific evidence is only one source of support for Creationism: "The discussion is primarily approached from the Biblical point of view, and assumes throughout that the Bible is the Word of God, divinely inspired and, therefore, completely reliable and authoritative on every subject with which it deals" (Morris 1972, v). But the appeal to scripture does not just provide extra support for Creationism, it seals the case: "We are forced to the conclusion, as Bible believing Christians, that the earth is really quite young after all, regardless of the contrary views of evolutionary geologists. This means then that all the uranium-lead measurements, the potassium-argon measurements and all similar measurements which have shown greater ages have somehow been misinterpreted" (Morris 1972, 89).

Of course, there is an obvious response to the charge that "Bible-believing Christians" must abandon evolutionary theory. Within Christianity, there is a long tradition of liberal interpretation of the Bible. Many Christian denominations, including some fundamentalist groups, do not insist that every sentence in the Bible should be read literally. They are aware of the difficulties involved in a literal reading, and, while they maintain that the Bible is a divinely inspired document, they are prepared for the possibility that, on a literal construction, the Genesis story is inaccurate.

Some passages in the Bible are obviously perplexing. There are lines in the Psalms suggesting that the earth is the center of the universe. Creationists feel that they are able to read *these* in a nonliteral way, because the Psalms are obviously poetry. The Creationist writings that we have read do not explicitly discuss the verse that gave the original Copernicans the most trouble: the description of how Joshua "commanded the sun to stand still" (Joshua 10 : 12–14). Watson claims that Galileo's claims conflicted only with "a few words of Bible poetry" (Watson 1976, 46); he does not deal with Galileo's (or the Church's?) major problem. (We understand from conversations with people who favor literal reading that Joshua's command should be regarded as requiring the cessation of *relative* motion and that this was simply expressed from a "human point of view.")

In any case, Creationists are not about to permit a liberal interpretation of the early chapters of Genesis. They regard this attitude as pernicious, or confused, or un-Christian (or all three). The Christian who would go down this path is solemnly warned that it "inevitably leads eventually to complete apostasy" (Morris 1974a, 247). Insofar as an argument is offered, it consists in pointing out that the early chapters of Genesis are written in the style of a historical narrative and that some New Testament writers (including Apostles) treat them as offering a historical narrative:

This type of Biblical exegesis [not treating Genesis as "true history"] is out of the question for any real believer in the Bible. It is the method of so-called "neo-orthodoxy," though it is neither new nor orthodox. It cuts out the foundation of the entire Biblical system when it expunges Genesis 1–11. The events of these chapters are recorded in simple narrative form, as though the writer or writers fully intended to record a series of straightforward historical facts; there is certainly no internal or exegetical reason for taking them in any other way. (Morris 1974a, 244)

Here the main thrust is clearly that someone who is prepared to abandon a literal reading of Genesis has no basis for taking any part of the Bible seriously. Watson makes the same point in a revealing way:

"Mommy, if God did not mean what He said, why did He not say what He *did* mean?"
The little girl's question highlights a problem that has faced every teacher of Genesis over the past hundred years. (Watson 1976, 11)

The only God worthy of mankind's trust and adoration is the God who can accurately describe the world's past, as a basis for predicting the world's future. (Watson 1976, 13)

So if one is to take the Bible as the Word of God, one must accept every word in it as literally true, including the beginning of Genesis. It is all very simple and convincing. Undoubtedly, this line has convinced many a sincere Christian to take a rosy view of Creation "science." But if the theological argument for Creationism is so straightforward, how can liberal theologians be so obtuse as to have missed it? As we have seen so many times before, matters are much more complex than the Creationists would like to think.

We begin with an obvious fact. The words in the Bible were written down by human beings. From the believer's perspective the writers were divinely inspired. But what exactly does this mean? There are two possible answers. First, the writers were simply *transcribers*. Their function was simply to write down words previously uttered or written by God—as, in the Mormon account, Joseph Smith had the function of transcribing and translating the tablets brought by the angel Moroni. Second, the writers were *authors*, composers of original words, who reported their experiences. If one accepts the first answer, then in the

Bible God has literally spoken. Hence it would be possible to ask—like Watson's little girl—why God did not say what He meant. However, given the second view, that question does not arise. The Genesis account of creation is not God's description of historical events. It is a narrative offered by human authors, people who may have been blessed with exceptional experiences, but who, nonetheless, were human.

We know of no compelling theological reason for preferring the first answer to the second. Certainly none is offered by the Creationists, whose arguments blur the distinction. Anybody who makes the distinction, and who accepts the second answer, can easily turn back the arguments that Morris and others offer. It is possible to accept the idea that the writers of Genesis *intended* to relate the true history of the origin of life, and to deny that their history is accurate. Nor do we discredit them by refusing to accept their narrative at face value. Human authors are fallible, and the works they write can mix important religious truth with factual error. (We can even accept the idea that the authors of the New Testament, themselves fallible humans, incorrectly believed that the Genesis narrative is accurate.) A religious person can accept evolutionary theory by elaborating further the idea of the Bible as *inspired* by God but *written* by humans. From this person's perspective, God originally created the universe, leaving it to evolve according to natural law. After the evolution of *Homo sapiens*, God chose to reveal Himself and some of His purposes to some of His creatures. The group of people in question, and some of their descendants (who may themselves also have had direct experiences of God), wrote books in which the important messages of the Creator are sometimes intertwined with highly inaccurate views about the workings of nature. As we discover things that they were in no position to know, we are able to identify places in which their primitive conceptions of the world overlay important truths. But to correct for these inadequacies should not diminish our respect for them or for the vital religious truths that their writings convey.

Yet it may seem that a residue of Watson's question remains. If the Bible does not record the true history of creation, why did the Creator allow the inspired people who wrote it to put forward so mistaken a picture of their origins? Why did God abstain from correcting their errors? These questions do not threaten the position of the religious evolutionist. Any theist will agree that God might have created wiser, more discerning creatures than *Homo sapiens*, or that God might have shared more knowledge with us. (Perhaps laypeople can even be excused for thinking that this is one message of the Book of Job.) For the religious evolutionist, the Bible contains information that God has chosen to transmit. There is no more reason to wonder why God did not clear up our confusion about the origin of the universe than to ask why God did not enlighten us about many issues on which we have only partial information. The believer maintains that God has made us aware of the important truths. But it has always been clear that we have not been told everything. Some things, like the true origin of life, we have to discover for ourselves.

Creationist arguments, such as they are, trade on linking the idea of the Bible as a divinely inspired document to the claim that every sentence it contains has to be taken literally. It is important to see that there is no tight connection here, and that the idea that religious evolutionists are compelled to deny the significance of the Bible is simply an attempt to bully Christians into accepting a doctrine for which there is not a shred of evidence.

Another Creationist tactic portrays the theory of evolution as severing the special relation between God and human beings. From an evolutionary point of view, we are a very recent development in the history of life. So why would an omniscient Creator choose so inefficient a method as evolution to achieve His purposes? As Morris asks, "If the goal of the evolutionary process was man, why did God take so long to get to the business at hand?" (Morris 1972, 73). A short answer to the question would be that God's purposes are not always apparent. Yet there is a serious worry, produced by reflection on the vastness of time, that the short answer does not touch. Evolutionary theory reveals *Homo sapiens* as simply one of the latest stages in a long process of development of living forms. There is no suggestion that humans are privileged, that our species is the focal point of divine concern. We have occupied the earth for a relatively short time, and, if we manage not to destroy it, our planet *may* one day provide a home for organisms who are our evolutionary descendants. Do not these evolutionary observations threaten religion by denying that there is a special relationship between humans and their Creator?

Freud claimed that humanity had suffered three great blows to its self-esteem. The first came when Copernicus declared that the earth is not the center of the universe and Galileo concluded that the universe is vastly larger than had hitherto been thought. The second was Darwin's proposal that living forms have evolved from common ancestors and that humans are descendants of animals who also gave rise to the contemporary apes. Freud thought that he himself had administered the third blow; by uncovering the workings of the unconscious, he had exposed us as far less rational than we had taken ourselves to be.

Whether Freud was correct to identify exactly these three incidents as critical in transforming our conception of ourselves, there is something acute in his characterization of them. They are indeed *blows to our self-esteem.* It is hard today to recapture the sense of uncertainty and loss that pervades some seventeenth-century writings. Donne remarks that "the new philosophy casts all in doubt," and Pascal reacts in horror to the infinite space of the universe. We have come to terms with the idea that our planet is not the physical center of a cozy Aristotelian cosmos. But, 125 years after Darwin, human vanity is still sometimes wounded by the thought of our kinship with the apes.

Yet, however much they may hurt our pride, Copernicanism and Darwinism are alike in that they do not question the possibility of a special relationship between God and humanity. Special relationships need not be exclusive. If, as some seventeenth-century cosmologists feared, the immensity of the universe suggests that we are not its only rational inhabitants, then that, in and of itself, need not diminish God's concern for us. If we are simply one among innumerable species who have occupied and will occupy our planet, then we are not compelled to conclude that we are not the focus of a special care. Religious belief is not threatened by the discovery that our part in the cosmic drama is confined to a single scene, unless we suppose that any object of divine concern will hog the stage throughout the entire play. Or, to switch metaphors, the religious believer should not behave like an eldest child who concludes that his parents no longer love him when he acquires a baby sister. To conceive of God as a Father, we need not suppose that we are His *only* children. It is enough to suppose that He cares for all His creatures and that He cares for each in ways that are appropriate to its abilities and needs.

Once we have adopted this perspective, we can see that the Creationist attempt to diagnose evolution as inefficient is based on a vain anthropocentrism. Evolutionary theory emphasizes our kinship with nonhuman animals and denies that we were created separately. But it does not interfere with the central Judaeo-Christian message that we are objects of special concern to the Creator. It simply denies us an exclusive right to that title.

In conclusion, let us look briefly at one of the more bizarre passages in the Creationist literature. Creationists could surely force Christians to choose between science and religion if they could successfully conclude that evolutionary theory is the invention of the Devil and specifically designed to overthrow the church of Christ. Henry Morris has suggested that this conclusion is literally true. Since 1859, many people have believed that Charles Darwin was the father of evolutionary theory. Morris thinks that they are quite wrong. Evolutionary theory was really devised by the Father of Lies Himself.

Creationists not only have their own "science." They also have their own "history of ideas," based on a surprising conspiracy theory. Morris claims that evolutionary theory is the outgrowth of an old movement, launched by the Forces of Darkness, that aims to overthrow Christianity. The movement includes an odd mixture of coconspirators: Immanuel Kant, for example, whose piety one might have taken to be beyond reproach, is accused of reviving "pagan philosophies" (Morris 1974b, 65). However, the main point of the history seems to be to trace evolutionary ideas to the early Greek atomists. After noting that the Greek atomists held that the order of the universe "arises out of a blind interplay of atoms" (Munitz 1957, 63; quoted in Morris 1974b, 66), Morris concludes triumphantly, "Modern evolutionary materialists are not so modern after all. Their system is essentially the same as the pre-Socratic Greek cosmology of 2500 years ago!" (Morris 1974b, 67). With this connection firmly in place, he goes on to appeal to an "unanswered" "classic work" (*The Two Babylons* by Alexander Hislop) that traces the early pagan ("evolutionary") philosophies back to Babylon. Now the stage is set. After assuming that "the Babylonian mysteries were originally established by Nimrod and his followers at Babel" (Morris 1974b, 73), Morris unmasks the horrible truth: "It therefore is a reasonable deduction, even though hardly capable of proof, that the entire monstrous complex was revealed to Nimrod at Babel by demonic influences, perhaps by Satan himself" (Morris 1974b, 74). Initially, Morris treats his hypothesis cautiously—claiming only that his account seems to be the best available treatment of the "known facts of the history of religions." But after a brief investigation of Satanic psychology, Morris becomes more confident. The discussion ends with a full-blown statement of the conspiracy theory: "[Satan] then brought about man's fall with the same deception ("ye shall be as gods") and the long sad history of the outworking of human unbelief as centered in the grand delusion of evolution has been the result" (Morris 1974b, 76).

We find it hard to believe that anybody—including Morris himself—accepts this shaggy-dog story. (One could make a far better case for tracing thermodynamics—specifically, the kinetic theory of heat—to the early Greek atomists. Are we then to think that this science too was inspired by the Devil?) We mention this curiosity only because it shows the lengths to which a prominent Creationist will go to attack the theory of evolution.

The ethics of survival

As the quotation beginning this chapter indicates, Creationists cannot be faulted for understating their concerns about the ethical implications of evolutionary theory. Still, they are not quite sure about how evolution is inimical to morality. Creationists claim both that evolutionary theory abolishes morality and that it fosters noxious ethical doctrines. Wysong maintains (Wysong 1976, 6) that the teaching of evolution promotes amorality: "Atoms have no morals, thus, if they are our progenitors, man is amoral. (I am not saying that an evolutionist is necessarily immoral, rather, I am saying that philosophically, logically, the term morality loses meaning in the context of true atheistic materialism.)" Yet, a page later, Wysong tells us that the fascism of Hitler and Mussolini was inspired by their reading of evolutionary writings. Presumably, if Wysong's argument about the atoms is correct, then Hitler and Mussolini must have imposed their own, antecedently held, moral doctrines on evolutionary theory. For if evolutionary theory denies meaning to morality, then evolution can have no moral principles of its own. One cannot complain both that a consistent theory denies that there are any moral principles and that it furnishes bad moral doctrines.

We shall examine the complaints separately. Wysong reaches his powerful conclusion that evolution portrays humans as amoral in one quick step: "Atoms have no morals, thus, if they are our progenitors, man is amoral" (Wysong 1976, 6). Atoms and molecules clearly lack moral qualities. We cannot praise a water molecule for its honesty or chide a segment of DNA for its intemperance. But these observations have no bearing on the question whether human conduct is subject to moral evaluation. There are innumerable cases in which an object, or a system of objects, is made up or developed from certain constituent or original parts and in which the resultant object has properties which the individual parts lack. It would be obvious folly to argue that people cannot live in brick houses, because, after all, people cannot live in individual bricks; or, to take another example, that apes cannot climb trees because the embryos from which they develop cannot climb trees. In the growth of individual organisms and in the formation of artifacts we recognize that features not present at early stages can be present later. Why should matters be different in the case of evolutionary development?

A clearheaded evolutionist will recognize that the history of life shows important turning points, after which some organisms have characteristics that no organism had before. For example, the first amphibians were the first vertebrates able to survive on land. Their achievement rested on a modification of structure; fins were transformed into rudimentary legs. An analogous story may be told in the case of moral properties and moral conduct. Presumably a number of characteristics underlie our capacities to develop moral virtues and moral practices, including the capacity to reason and to reflect on our actions. Bypassing what are in fact important questions, let us assume that our apelike ancestors and contemporary apes both lack all or most of the capacities needed for moral action. Contemporary apes and our apelike ancestors would then be amoral beings. But what does that say about us? The fact that organisms with certain features evolved out of organisms that lacked those features does not deprive the descendant organisms of the crucial features. Birds (with wings) evolved out of wingless reptiles; but, for all that, birds still have wings. Ever since Darwin, some people have shown an unaccountable tendency to read the claim that we evolved from apelike ancestors by fastening on the *apishness* of our ancestors and gliding past the important point that we have *evolved from* these creatures. It would be incorrect to speak of our *evolution* from apelike organisms unless, besides many similarities, there were significant differences between us and them. If they lacked the wherewithal to develop moral virtues, moral codes, and, hence, moral behavior, then, unless we are very mistaken about ourselves, this is a significant area of difference.

So much for Wysong's muddles about molecules. We now turn to the other Creationist complaint, the charge that the scientific theory of evolution purveys an immoral moral philosophy. Morris devotes large sections of *The Troubled Waters of Evolution* to the theme that evolution leads to innumerable forms of wickedness. Here are some sample passages:

The philosophies of Karl Marx and Friedrich Nietzsche—the forerunners of Stalin and Hitler—have been particularly baleful in their effects. Both were dedicated evolutionists. (Morris 1974b, 33)

It is clear that the inexorable logic of evolutionary reasoning leads directly to the conclusion that war and struggle is the chief good, leading to evolutionary advance. (Morris 1974b, 36)

One of the frightening things about modern evolutionists and sociologists is that they have come to believe they should control future evolution. This they propose to do by genetic manipulations of various sorts, by control of births and possibly of deaths, by an intellectual elite who will decide who is fit to have children and what kinds of babies are desirable, and by state enforcement of their decisions. (Morris 1974b, 46–47)

All the great political and economic movements — whether communism, economic imperialism, centralized capitalism, racism, or others — were all eager to grab up Darwinism as justification for their particular brands of man's basic self-centered struggle-and-survival ethic. (Morris 1974b, 59)

As the 19th century scientists were converted to evolution, they were thus also convinced of racism. They were certain that the white race was superior to other races, and the reason for this superiority was to be found in Darwinian theory. (Morris 1974b, 164)

Furthermore, since animals are indiscriminate with regard to partners in mating and, since men and women are believed to have evolved from animals, then why shouldn't we live like animals? (Morris 1974b, 167–168)

We would emphasize that this is only a *small* sample. There are many other ills that Morris and his fellows intend to trace to evolutionary ideas (see Morris 1974b, 25–62 passim, 154–184). But it should be clear what the basic evolutionary maxim is supposed to be: Make Love *and* War.

Many of these charges invite an obvious response. Various people have appealed to the theory of evolution to lend respectability to their appalling moral views. The nineteenth-century Social Darwinists, the proponents of some forms of racism, Hitler, the eugenicists, and a host of others have all tried to enlist the theory of evolution in support of their causes. But this fact says very little about evolutionary theory itself. Virtually any morally neutral, or even morally good, doctrine can be misused for evil purposes. (A striking illustration will be given later in the chapter.) Certainly any scientific theory can be thus abused. If somebody espouses the peculiar belief that the higher the vocal register a person has the less worthy that person is, then it is possible to use acoustical and physiological findings to infer that women are generally inferior to men. This conclusion does not reveal that acoustics and physiology are sexist sciences. It simply shows that if people know the moral conclusion they wish to reach, then they can use carefully chosen bits of science, *plus* evaluative claims of their own, to provide a "scientific" defense of their views.

The racism of nineteenth-century evolutionists, to which Morris alludes, provides a clear illustration. A number of nineteenth-century biologists shared two beliefs. They held that blacks are inferior to whites and that humans evolved from apelike ancestors. This inspired them to look for ways of measuring the extent to which a person has departed from the original apelike form, ways that were carefully adjusted, so that, *in conjunction with the evaluative claim that the worth of a person is proportional to the departure from apelike form*, they could deduce that whites are superior. Notice that the evaluative claim is no part of evolutionary theory; that theory does not pronounce on the relative worth of organisms. The evaluative claim, with its commitment to a carefully chosen method of measuring distance from the apes, is accepted simply because it permits a spurious "explanation" of an antecedently accepted racist conclusion. Evolution did not supply the racist doctrine. It merely played a mediating role, linking two racist claims. (For more details on the example, see Gould 1981a.)

Morris is ready for this obvious defense. He tries to block it by denying the possibility that immoral doctrines result from perversions of evolutionary theory. Toward the end of *The Troubled Waters of Evolution*, he sums up his complaints:

The evolutionary philosophy is the intellectual basis of all anti-theistic systems. It served Hitler as the rationale for Nazism and Marx as the supposed scientific basis for communism. It is the basis of the various modern methods of psychology and sociology that treat man merely as a higher animal and which have led to the mis-named "new morality" and ethical relativism. It has provided the pseudo-scientific rationale for racism and military aggression. Its whole effect on the world and mankind has been harmful and degrading. Jesus said: "A good tree cannot bring forth evil fruit" (Matthew 7 : 18). The evil fruit of the evolutionary philosophy is evidence enough of its evil roots. (Morris 1974b, 186)

Even if he can use this stick to beat evolutionary theory, Morris has made a desperately unwise choice of weapons. The most popular doctrine for use in rationalizing evil and immoral actions has surely been Christianity. There is a long record of brutalities and atrocities perpetrated in the name of Christ: the Crusades, the persecution of the Huguenots, periodic waves of anti-Semitism, sporadic witch burnings, the Inquisition, 300 years of Irish "troubles"; the list could go on and on. Add to this the explicit racism of some *contemporary* Christian sects, the repressive moral doctrines imposed by the Church at many times in the past, the denials of justice and basic human rights in the name of the "divine right of Christian princes." We could easily parody

Morris's conclusion: The evil fruit of Christianity is evidence enough of its evil roots.

Yet although the Christian church has a checkered history, it is evident that Christians can claim—quite justifiably—that the evils result from perversions of religious doctrine: Evil or misguided men have twisted the Gospel to evil ends. Morris himself seems to appreciate this point: "If certain Christian writers have interpreted the Bible in a racist framework, the error is in the interpretation, not in the Bible itself" (Morris 1974b, 162). But if charity ought to be extended to Christian doctrine, then it is equally appropriate for evolutionary theory. Both the Bible and evolutionary theory can be misread and their principles abused. Hitler's anti-Semitism is no more a fruit of evolutionary theory than it is of Christianity.

(Like Morris's list of evil abuses of evolution, our own survey of perversions of Christianity may leave a misleading impression. Psychological studies have shown that, even when they are explicitly retracted, charges have a tendency to stick. When subjects are told that a certain object has unpleasant properties and when they are later informed that the original descriptions were incorrect, the first impression still lingers. The subjects retain the idea that the object in question is bad. Of course, lawyers knew about this long before the psychologists. The tactic of introducing suggestions that later have to be "stricken from the record" exploits it. Hence, even though Morris's criticisms of evolution and our parallel about Christianity are illegitimate for the same reason, they may encourage the attitude that both doctrines are evil. That attitude is mistaken. The fact that both doctrines can be perverted should not lead us to think that either encourages immoral actions. The same psychological studies have also shown that the only way to counter the effects of such misleading information seems to be to tell people about the phenomenon—hence the present paragraph.)

Morris's attempt to brand evolutionary theory as responsible for the actions of people of whom he thinks his readers will disapprove turns out to be a fiasco. However, the passages we have quoted present another charge. Evolutionary theory portrays nature as the scene of a struggle to survive and reproduce in which rapacity and sexual promiscuity may be rewarded (by greater representation in subsequent generations). Morris draws the moral that evolutionary theory *commends* whatever actions and stratagems are useful in the struggle. Thus he suggests that evolution supplies its own ethical principles, directing us to kill the competition and to engage in indiscriminate sex.

There are errors of fact and of judgment here. As writers from Kropotkin to the contemporary sociobiologists have observed, cooperation and "altruistic" behavior can sometimes be advantageous. Moreover, not all animals "mate indiscriminately"; some animals pair far more faithfully than many humans. But even if Morris were right in his facts, even if cruelty and sexual promiscuity were natural to all animals, including humans, what ethical conclusions would follow? To derive a maxim for our conduct we need a further ethical assumption: We *ought* to do what comes naturally.

Words like *natural* and *fit* are very deceptive. Despite the practices of Madison Avenue, *natural* is not an all-purpose word of commendation. (This ploy may have reached its height, or nadir, when one company advertised its pesticide as "all natural, contains no artificial chemicals.") Some natural processes have very undesirable effects. Rickets is a natural result of too little vitamin D and/or calcium in childhood. However, it is hard to believe that anyone thinks that it is a good thing that malnourished children develop rickets. To say that a characteristic is natural is to say only that it is the result that would follow in the absence of some type of intervention; it is not to say that the trait is good or that intervention would be misplaced.

The ideas of fitness, and of the "survival of the fittest," are probably even more misleading. It is well to recall the *precise* definitions of evolutionary theory. A characteristic promotes fitness if, given a particular environment of supply and competition, that trait will increase the probability of genetic representation in future generations. But what does that really say about the trait? To begin with, it is important to realize that, to a degree unrivaled by other animals, human beings are capable of shaping the environment to their own designs. So if we discover that a morally praiseworthy characteristic lowers fitness in a particular environment, one obvious response would be to try to alter the environment.

Let us now look at the worst case. Let us suppose that an evolutionary study of a population of humans reveals that genes present in that population will be lost unless the humans in question engage in acts of aggression and pursue lives of wanton depravity—and that, for some reason, those people are powerless to change their environment. What does the theory of evolution imply that these unfortunates should do? Even in these circumstances, it does not imply that aggression and depravity are morally correct. The theory says only that these are the only means by which the threatened genes can be saved. It does not say that this is a desirable end. It does not say that this end is sufficient to justify these means. No strictly *evolutionary* study will make pronouncements on those questions. If the people decide that violence and wantonness are the best available courses of action, that will be because they assign the highest value to survival, or at least a higher value to survival than to some ethical practices. Evolutionary theory recommends an "ethics of survival" only to those people who come to the theory already convinced of the intrinsic and overwhelming worth of survival.

Besides the charges already considered, Morris claims to find yet another skeleton in the evolutionist's closet: ethical relativism. Creationists frequently cite the following observation by A. G. Motulsky: "An ethical system that bases its premises on absolute pronouncements will not usually be acceptable to those who view human nature by evolutionary criteria" (Motulsky 1974, 654; cited in Morris 1975, 13; Wysong 1976, 6; Morris 1974b, 34, 183). Perhaps some people sense a linkage between these two views, although philosophers decidedly do not. Most twentieth-century philosophers are staunch supporters of the theory of evolution. Yet much of the work of twentieth-century ethics has been devoted to combating one or another form of ethical relativism. Why should anyone think that the theory of evolution implies that moral values are only relative? Evolutionary theory may reveal that the strategies for promoting our species have changed during the course of its historical development. Different properties and forms of behavior may have been appropriate for successful competition at different times. But that does not imply that different forms of behavior were morally correct, or even morally valuable, at different times. That moral assessment follows only if we assume an additional moral premise: A property or form of behavior is good if it promotes survival (regardless of its other features). As we have already seen, people learn the ethics of survival from evolutionary theory only when they view the theory in light of antecedently held beliefs about the value of survival. To take a specific example, an evolutionary biologist could consistently hold that it is absolutely impermissible for humans to eat the flesh of other mammals—even though he recognizes that our progenitors engaged in the forbidden activity and that their doing so made it possible for us to exist and to evaluate their actions. What is evolutionarily useful, or even necessary, may not be morally correct.

Who's afraid of evolutionary theory?

Thomas Henry Huxley was a man of many accomplishments. Not only was he able to demolish Bishop Wilberforce with a timely piece of repartee. He was also one of the first people to see clearly that the theory of evolution does not provide a simple recipe for moral and social reform. We have been defending one of Huxley's most celebrated insights. It may be profitable to study the evolution of ethics, but that does not imply that there is an ethics of evolution.

Huxley appreciated the complexity of moral thinking. Constructing a system of ethics is not a task for some rainy afternoon when one has nothing better to do. To deliberate seriously about what ought to be done involves deep and careful reflection on a broad range of considerations. The point is easily appreciated when we recall our own difficult decisions or when we try to empathize with the deliberations of others. Consider, for example, the predicament of a young man who is called to serve in a war he perceives as immoral. In making a decision about what he should do, he must balance different demands. As a citizen he has an obligation to obey the state. Yet his recognition of the evil purposes of the war directs him to disobey. On the other hand, he realizes that if he does not serve, somebody else will be forced to go in his place. Moreover, he will appreciate the need to scrutinize his own motives; is his abhorrence of the war fed by cowardice or disinclination to disrupt his civilian existence? The process of moral reflection is extremely complex.

Like other sciences that study humans and our relation to the rest of the world, evolutionary theory can yield information that is relevant to our moral reflections. As we understand more about ourselves, the intricate chains of reasoning that underlie responsible individual and social decisions will be affected. But moral arguments are too intricate for scientific findings, in and of themselves, to produce a full-blown moral theory. A major new theory about human characteristics or human behavior may alter some applications of central moral principles, but it is unlikely to force a revision of the basic tenets of the system of ethics we have developed over the last 2,000 years. Only someone who does not appreciate the complexity of moral thought will believe that science can resolve controversial issues at a stroke or that the science of evolution could "refute" a broadly based ethical system.

If the issues were not so serious, laughter would be the appropriate response to the Creationists' quick indictment of evolution: We learn of our kinship with other animals, so we turn savage and promiscuous, tear down our social institutions, and abandon our ordinary attitudes to personal relations; all the ills of the present age descend from Darwin; high crime rates, perversion, prostitution, greed, communism, nazism—anything the intended audience is assumed to dislike—all flow from this single source. Still, it does seem possible to doubt that the Happy Hooker keeps a copy of *The Origin of Species* tucked under her pillow or that the average mugger draws inspiration from a careful study of the chapter on "The Struggle for Existence." No doubt new slings and arrows are in store for Darwin. An unending flow of evil consequences must somehow be traced to the theory of evolution so that it becomes "easier" to choose between evolution and Creationist fundamentalism. Perhaps the real cause of high interest rates, record deficits, and increased unemployment will turn out to be a "survivalist ethic." In the gothic fantasies of Creationists, all such connections are possible.

The Creationists are by no means the first to play on fears about what scientific inquiry will disclose. Anxieties about ourselves endure. If our proper study is indeed the study of humankind, then it has seemed—and still seems—to many that the study is dangerous. Perhaps we shall find out that we were not what we took ourselves to be. But if the historical development of science has indeed sometimes pricked our vanity, it has not plunged us into an abyss of immorality. Arguably, it has liberated us from misconceptions, and thereby aided us in our moral progress.

The theory of evolution explains to us what our ancestry has been. It does not explain away our worth. Why should we be afraid to learn more about what we are? As T. S. Eliot wrote in "Little Gidding,"

We shall not cease from exploration
And the end of all our exploring
Will be to arrive where we started
And know the place for the first time.

Suggestions for Further Reading

Chapter 1

An excellent introduction to evolutionary theory is provided in John Maynard Smith's *The Theory of Evolution* (London: Penguin, 1975). More advanced, but extremely lucid and comprehensive, are *Evolution* by T. Dobzhansky, F. J. Ayala, G. L. Stebbins, and J. Valentine (San Francisco: Freeman, 1977) and *Evolutionary Biology* by D. J. Futuyma (Sunderland, MA: Sinauer, 1979). *Genetics* by M. W. Strickberger (New York: Macmillan, 1976) offers a good introduction to Mendelian genetics, chromosomal genetics, and molecular genetics. H. L. K. Whitehouse's *Towards an Understanding of the Mechanism of Heredity* offers an interesting historical approach to genetics. The mathematical parts of population genetics are explained with great clarity in *A Primer of Population Biology* by E. O. Wilson and W. Bossert (Sunderland: Sinauer, 1971). An authoritative discussion of population genetics can be found in R. C. Lewontin's lucid book *The Genetic Basis of Evolutionary Change* (New York: Columbia, 1974).

Chapter 2

It is difficult to find a good way in to contemporary philosophy of science. Carl G. Hempel's *Philosophy of Natural Science* (Englewood Cliffs: Prentice-Hall, 1966) is admirably clear, but its treatment of some issues (including topics discussed in this book) is somewhat dated. Thomas Kuhn's *The Structure of Scientific Revolutions* (Chicago: University of Chicago Press, 1970) is an extremely stimulating discussion of scientific change, but it has been known to mislead the unwary reader. There are provocative— but controversial—accounts of scientific methodology in Imre Lakatos's *The Methodology of Scientific Research Programmes* (Cambridge: Cambridge University Press, 1979) and L. Laudan's *Progress and Its Problems* (Berkeley: University of California Press, 1977). The point of view of the present book has affinities with more technical treatments of the methods of science in Clark Glymour's *Theory and Evidence* (Princeton: Princeton University Press, 1980); Sylvain Bromberger's "Science and the Forms of Ignorance," in M. Mandelbaum (ed.), *Theory and Observation in Science* (Baltimore: Johns Hopkins University Press, 1971); and my own "Explanatory Unification," *Philosophy of Science* 1981(48), 507–531.

Chapter 3

Discussions of some methodological issues in evolutionary biology, and of the tautology objection, are given in David Hull, *Philosophy of Biological Science* (Englewood Cliffs:

Prentice-Hall, 1974) and Michael Ruse, *Philosophy of Biology* (London: Hutchinson, 1974). Hull's anthology *Darwin and His Critics* is a valuable collection, containing some very interesting presentations of objections to evolution. Ernst Mayr's *Populations, Species and Evolution* (Cambridge: Harvard University Press, 1970) is an extremely clear introduction to contemporary views about speciation, and George C. Williams's *Adaptation and Natural Selection* (Princeton: Princeton University Press, 1966) offers an incisive critique of concepts of evolutionary progress.

Chapter 4

The issue of the randomness of mutations is addressed very well in Futuyma's *Evolutionary Biology*. Philip Morse's *Thermal Physics* (New York: Benjamin, 1974) is a thorough survey of classical and contemporary thermodynamics. *Introduction to the Principles of Paleontology* by David Raup and Steven Stanley is an excellent general study of the fossil record and what it can reveal. A. S. Romer's *Vertebrate Paleontology* (Chicago: University of Chicago Press, 1966) is an encyclopedic work on the evolution of the vertebrates. E. H. Colbert's *Evolution of the Vertebrates* (New York: Wiley, 1980) is less detailed, but easier reading. Information about early mammals can be found in *Mesozoic Mammals*, edited by L. Lillegraven, Z. Kielan-Jaworowska, and W. Clemens (Berkeley: University of California Press, 1979). The Paluxy finds are discussed in C. G. Weber, "Paluxy Man—the Creationist Piltdown," *Creation/Evolution* 1981(6), 16–22.

Chapter 5

The best presentation of contemporary Creationism is Henry Morris's *Scientific Creationism* (San Diego: Creation-Life Publishers, 1974). *The Genesis Flood*, by Morris and John Whitcomb (Grand Rapids: Baker, 1978), has more detail on some issues, but mixes scientific and religious evidence in a way that Creationists no longer favor (at least for their "scientific" works). C. C. Gillispie's *Genesis and Geology* is a very good account of the debates about Flood Geology in the early nineteenth century. Stephen Jay Gould's collections *Ever Since Darwin* (New York: Norton, 1977) and *The Panda's Thumb* (New York: Norton, 1980) contain a number of essays relevant to the themes of the chapter. For the theory of punctuated equilibria, the *locus classicus* is N. Eldredge and S. J. Gould, "Punctuated Equilibria: An Alternative to Phyletic Gradualism," in T. J. M. Schopf (ed.), *Models in Paleobiology* (San Francisco: Freeman, 1972). Finally, D. Eicher's *Geological Time* (Englewood Cliffs: Prentice-Hall, 1968) provides a good introduction to dating techniques, including methods of radiometric dating.

Chapters 6 and 7

These chapters are relatively nontechnical and do not draw on a body of specialized literature. Those who are interested in the relation between evolution and ethics may wish to look at Richard Brandt's *Ethical Theory* (Englewood Cliffs: Prentice-Hall, 1959), a book that provides comprehensive discussion of various approaches to the foundations of ethics. A major, original contemporary work in ethics is John Rawls's *A Theory of Justice* (Cambridge: Harvard University Press, 1971).

References

Asimov, I. 1979. "In the Game of Energy and Thermodynamics, You Can't Even Break Even." *Journal of the Smithsonian Institution.*

Barnes, T. 1973. *Origin and Destiny of the Earth's Magnetic Field.* San Diego: Creation-Life Publishers.

Bethell, T. 1976. "Darwin's Mistake." *Harper's* (February issue).

Birch, L. C., and Ehrlich, P. R. 1967. "Evolutionary History and Population Biology." *Nature* 214, 349–352.

Brush, S. G. 1982. "Finding the Age of the Earth: By Physics or by Faith?" *Journal of Geological Education* 30, 34–58.

Campbell, B. 1974. *Human Evolution.* Chicago: Aldine.

Clark, M. E. 1976. *Our Amazing Circulatory System . . . By Chance or Creation?* San Diego: Creation-Life Publishers.

Clark, W. E. le Gros 1964. *Fossil Evidence for Human Evolution.* Chicago: University of Chicago Press.

Clark, W. E. le Gros 1971. *The Antecedents of Man.* Chicago: University of Chicago Press.

Colbert, E. H. 1980. *Evolution of the Vertebrates.* New York: Wiley.

Creation-Life Publishers. 1977. *21 Scientists Who Believe in Creation.* San Diego: Creation-Life Publishers.

Crompton, A. W., and Jenkins, F. A. 1979. "Origin of Mammals." In J. A. Lillegraven, Z. Kielan-Jaworowska, and W. A. Clemens (eds.), *Mesozoic Mammals.* Berkeley: University of California Press.

Crompton, A. W., and Parker, P. 1978. "Evolution of the Mammalian Masticatory Apparatus." *American Scientist* 66, 192–201.

Dalrymple, G. B. forthcoming. *Radiometric Dating, Geologic Time, and the Age of the Earth: A Reply to "Scientific" Creationism.* Typescript U.S. Geological Survey.

Darwin, C. 1859/Mayr, E. 1964. *The Origin of Species.* Facsimile of the first edition. Cambridge: Harvard University Press.

Darwin, C. 1876/Barlow, N. 1969. *The Autobiography of Charles Darwin.* New York: Norton.

Dobzhansky, T. 1973. "Nothing in Biology Makes Sense Except in the Light of Evolution." *American Biology Teacher* 35, 125–129.

Dobzhansky, T., Ayala, F. J., Stebbins, G.L., and Valentine, J. W. 1977. *Evolution.* San Francisco: Freeman.

Eicher, D. L. 1968. *Geologic Time.* Englewood Cliffs: Prentice-Hall.

Eldredge, N., and Gould, S. J. 1972. "Punctuated Equilibria: An Alternative to Phyletic Gradualism." In T. J. M. Schopf (ed.), *Models in Paleobiology.* San Francisco: Freeman.

Faul, H. 1966. *Ages of Rocks, Planets, and Stars.* New York: McGraw-Hill.

Futuyma, D. J. 1979. *Evolutionary Biology.* Sunderland, MA: Sinauer.

Gillispie, C. C. 1951. *Genesis and Geology.* New York: Harper.

Gish, D. T. 1979. *Evolution? The Fossils Say No!* San Diego: Creation-Life Publishers.

Gish, D. T. 1981. *Acts, Facts and Impacts* (December issue).

Gould, S. J. 1980. *The Panda's Thumb.* New York: Norton.

Gould, S. J. 1981a. *The Mismeasure of Man.* New York: Norton.

Gould, S. J. 1981b. "Evolution as Fact and Theory." *Discover* 2, 34–37.

Gould, S. J., and Eldredge, N. 1977. "Punctuated Equilibria: The Tempo and Mode of Evolution Reconsidered." *Paleobiology* 3, 115–151.

Hallam, A. 1973. *A Revolution in the Earth Sciences.* New York: Oxford University Press.

Hefley, J. C. 1979. *Are Textbooks Harming Your Children?* Milford, MI: Mott Media.

Hempel, C. G. 1951. "Problems and Changes in the Empiricist Criterion of Meaning." In *Aspects of Scientific Explanation.* Glencoe: The Free Press, 1965.

Hiebert, H. 1979. *Evolution: Its Collapse in View?* Alberta: Horizon Books.

Hull, D. L. 1974. *Darwin and His Critics.* Cambridge: Harvard University Press.

Jenkin, F. 1867. "Review of *The Origin of Species.*" *The North British Review.* Reprinted in Hull 1974.

Kuhn, T. S. 1970. *The Structure of Scientific Revolutions.* Chicago: University of Chicago Press.

Kurtén, B. 1971. *The Age of Mammals.* London: Weidenfeld.

Leibniz, G. F. W. 1686/1979. *Discourse on Metaphysics.* LaSalle: Open Court.

Lovejoy, A. O. 1942. *The Great Chain of Being.* Cambridge: Harvard University Press.

Macbeth, N. 1971. *Darwin Retried.* Boston: Gambit.

Matthews, L. H. 1971. Introduction to *The Origin of Species.* London: Dent.

Maynard Smith, J. 1978. *The Evolution of Sex.* Cambridge: Cambridge University Press.

Mayr, E. 1959. "The Emergence of Evolutionary Novelties." In Mayr 1976.

Mayr, E. 1963. *Animal Species and Evolution.* Cambridge: Harvard University Press.

Mayr, E. 1970. *Populations, Species, and Evolution.* Cambridge: Harvard University Press.

Mayr, E. 1976. *Evolution and the Diversity of Life.* Cambridge: Harvard University Press.

Moore, J. 1974. *Should Evolution Be Taught?* San Diego: Creation-Life Publishers.

Morris, H. M. 1972. *The Remarkable Birth of Planet Earth*. San Diego: Creation-Life Publishers.

Morris, H. M. 1974a. *Scientific Creationism* (general edition). San Diego: Creation-Life Publishers. [With the exception of the passages cited from pp. iii and 244 (the foreword and final chapter), all quotations from *Scientific Creationism* will be found verbatim on the same pages of the "public school" edition.]

Morris, H. M. 1974b. *The Troubled Waters of Evolution*. San Diego: Creation-Life Publishers.

Morris, H. M. 1975. *Introducing Creationism in the Public Schools*. San Diego: Creation-Life Publishers.

Morse, P. 1974. *Thermal Physics*. New York: Benjamin.

Motulsky, A. G. 1974. "Brave New World?" *Science* 185, 653–663.

Munitz, M. K. 1957. *Theories of the Universe*. Glencoe: The Free Press.

Nelkin, D. 1977. *Science Textbook Controversies and the Politics of Equal Time*. Cambridge: MIT Press.

Newton, I. 1687/Motte-Cajori 1960. *The Mathematical Principles of Natural Philosophy*. Berkeley: University of California Press.

Oxnard, C. 1975. "The Place of the Australopithecines in Human Evolution: Grounds for Doubt?" *Nature* 258, 386–394.

Peters, R. H. 1976. "Tautology in Evolution and Ecology." *American Naturalist* 110, 1–12.

Popper, K. R. 1959. *The Logic of Scientific Discovery*. London: Hutchinson.

Popper, K. R. 1963. *Conjectures and Refutations*. New York: Harper.

Quine, W. V. O. 1952. "Two Dogmas of Empiricism." Reprinted in *From a Logical Point of View*. Cambridge: Harvard University Press.

Raup, D. M. 1979. "Conflicts between Darwin and Paleontology." *Field Museum of Natural History Bulletin* 50, 22–29.

Raup, D. M., and Stanley, S. 1978. *Principles of Paleontology*. San Francisco: Freeman.

Romer, A. S. 1966. *Vertebrate Paleontology*. Chicago: University of Chicago Press.

Schramm, D. N. 1974. "The Age of the Elements." *Scientific American* 230, 69–77.

Sedgwick, A. 1831. *Proceedings of the Geological Society of London*, 313–314.

Simon, H. 1968. *The Sciences of the Artificial*. Cambridge: MIT Press.

Simpson, G. G. 1953. *The Major Features of Evolution*. New York: Columbia University Press.

Slusher, H. S., and Gamwell, T. P. 1978. *The Age of the Earth*. San Diego: Creation-Life Publishers.

Vaughan, T. A. 1978. *Mammalogy*. Philadelphia: Saunders.

Watson, D. C. C. 1976. *The Great Brain Robbery*. Chicago: Moody Press.

Whitcomb, J., and Morris, H. M. 1961. *The Genesis Flood*. Grand Rapids: Baker.

White, M. J. D. 1977. *Modes of Speciation.* San Francisco: Freeman.

Wilder-Smith, A. E. 1970. *The Creation of Life.* San Diego: Creation-Life Publishers.

Wilder-Smith, A. E. 1981. *The Natural Sciences Know Nothing of Evolution.* San Diego: Creation-Life Publishers.

Williams, G. C. 1966. *Adaptation and Natural Selection.* Princeton: Princeton University Press.

Williams, G. C. 1975. *Sex and Evolution.* Princeton: Princeton University Press.

Wilson, E. O., and Bossert, W. H. 1971. *A Primer of Population Biology.* Sunderland, MA: Sinauer.

Wysong, R. L. 1976. *The Creation-Evolution Controversy.* Midland, MI: Inquiry Press.

Index

The index contains three types of entries. There are technical terms, whose page references are given in italics to show where the meanings of the terms are explained. There are names of people (places, kinds of organisms, and so forth), whose page references include all pertinent discussions of the people (places and organisms) in question. Finally, there are entries intended to help the reader locate quickly discussions of particular aspects of the creation-evolution controversy; for example, under the entries *Evolutionary theory* and *"Scientific" Creationism* there are references to specific evaluations of these doctrines.